# Polarimetric SAR Techniques and Applications

## Special Issue Editors

Carlos López-Martínez
Juan Manuel Lopez-Sanchez

MDPI • Basel • Beijing • Wuhan • Barcelona • Belgrade

MDPI

*Special Issue Editors*

Carlos López-Martínez
Universitat Politècnica de Catalunya
Spain

Juan Manuel Lopez-Sanchez
Universidad de Alicante
Spain

*Editorial Office*
MDPI AG
St. Alban-Anlage 66
Basel, Switzerland

This edition is a reprint of the Special Issue published online in the open access journal *Applied Sciences* (ISSN 2076-3417) in 2017 (available at: http://www.mdpi.com/journal/applsci/special_issues/SAR).

For citation purposes, cite each article independently as indicated on the article page online and as indicated below:

Author 1; Author 2. Article title. *Journal Name* **Year**, *Article number*, page range.

**First Edition 2017**

**ISBN 978-3-03842-616-5 (Pbk)**
**ISBN 978-3-03842-617-2 (PDF)**

# Table of Contents

# About the Special Issue Editors

**Carlos López-Martínez**, Ph.D. He received his MSc. degree in electrical engineering and his Ph.D. degree from the Universitat Politècnica de Catalunya, Barcelona, Spain, in 1999 and 2003, respectively. From October 2000 to March 2002, he was with the Frequency and Radar Systems Department, HR, German Aerospace Center, DLR, Oberpfaffenhofen, Germany. From June 2003 to December 2005, he was with the Image and Remote Sensing Group—SAPHIR Team, in the Institute of Electronics and Telecommunications of Rennes (I.E.T.R.—CNRS UMR 6164), Rennes, France. In January 2006, he joined the Universitat Politècnica de Catalunya as a Ramón-y-Cajal researcher, Barcelona, Spain, where he was associate professor in the area of remote sensing and microwave technology until April 2017. Since May 2017, he has been responsible for the Remote Sensing and Eco-Hydrological Modelling group in the Luxembourg Institute of Science and Technology, Luxembourg. His research interests include SAR and multidimensional SAR, radar polarimetry, physical parameter inversion, digital signal processing, estimation theory and harmonic analysis. He is associate editor of IEEE Journal of Selected Topics in Applied Earth Observations and Remote Sensing and he served as guest editor of the EURASIP Journal on Advances in Signal Processing. He has organized different invited sessions at international conferences on radar and SAR polarimetry. He has presented advanced courses and seminars on radar polarimetry to a wide range of organizations and events. Dr. López-Martínez has authored or co-authored more than 100 articles in journals, books, and conference proceedings in the Radar Remote Sensing and image analysis literature. He received the Student Prize Paper Award at the EUSAR 2002 Conference and co-authored the paper awarded with the First Place Student Paper Award at the EUSAR 2012 Conference. He has also received the IEEE-GRSS 2013 GOLD Early Career Award.

**Juan Manuel Lopez-Sanchez**, Ph.D. He received his Ingeniero (M.S.) and Doctor Ingeniero (Ph.D.) degrees in Telecommunication Engineering from the Technical University of Valencia (UPV), Valencia, Spain, in 1996 and 2000, respectively. From 1998 to 1999 he worked as a predoctoral grantholder at the Space Applications Institute, Joint Research. Centre of the European Commission, Ispra, Italy. Since 2000, he has led the Signals, Systems and Telecommunication Group of the University of Alicante, Spain, where he has been full professor since November 2011. His main research interests include microwave remote sensing for inversion of biophysical parameters, polarimetric and interferometric techniques, SAR imaging algorithms, and applications of radar remote sensing in agriculture and geophysics. In 2001, Dr. Lopez-Sanchez received the Indra Award for the best Ph.D. thesis about radar in Spain. From 2006 to 2012, he was the Chair of the Spanish Chapter of the IEEE Geoscience and Remote Sensing Society. He has coauthored more than 60 papers in refereed journals and more than 110 papers and presentations at international conferences and symposia.

*applied sciences*

MDPI

*Editorial*

# Special Issue on Polarimetric SAR Techniques and Applications

Carlos Lopez-Martinez [1,*] and Juan M. Lopez-Sanchez [2,*]

[1]  ERIN Department, Luxembourg Institute of Science and Technology (LIST), L-4422 Belval, Luxembourg
[2]  DFISTS–IUII, University of Alicante, P.O. Box 99, 03080 Alicante, Spain
*   Correspondence: carlos.lopez@list.lu (C.L.-M.); juanma.lopez@ua.es (J.M.L.-S.);
    Tel.: +352-275-888-5056 (C.L.-M.); +34-965-909-597 (J.M.L.-S.)

Received: 26 July 2017; Accepted: 27 July 2017; Published: 28 July 2017

## 1. Introduction

Synthetic Aperture Radar (SAR) polarimetry is an active and fruitful field of research in Earth observation. Polarimetry provides sensitivity to the soil moisture, as well as to the structural and geometric properties of the targets under observation, allowing a more accurate identification and classification than with non-polarimetric data. Moreover, the increasing number of spaceborne SAR systems equipped with polarimetric capabilities, as well as future planned missions, enables the advance in this research field at all levels, from theory and physical modeling to final applications.

## 2. Polarimetric SAR: Techniques and Applications

This special issue was introduced to collect the latest research on relevant aspects of SAR polarimetry, to present state-of-the-art developments, and to show the current and futures challenges of SAR polarimetry with the availability of new sources of data. Therefore, this special issue also places an emphasis on studies for the exploitation of data provided by the new polarimetric spaceborne SAR sensors, which include additional frequency bands, interferometric capabilities, enlarged spatial coverage, high spatial resolution, and/or shorter revisit times. There were 17 papers submitted to this special issue, and nine papers were accepted (i.e., 50% acceptance rate). The published papers can be grouped into three main topics: polarimetric data classification, SAR polarimetry applications, and polarimetric SAR interferometry (PolInSAR).

From all the papers accepted in this special issue, three of them focused on polarimetric SAR data classification, covering different techniques, as well as different types of land surfaces. The first paper, authored by X. Wang, Z. Cao, Y. Ding, and J. Feng, introduces a composite kernel method, based on a Support Vector Machine classification approach [1]. The contribution of this paper is that data classification is based on a weighted combination of both polarimetric information and spatial characteristics derived from the Span image. As demonstrated by the authors, the introduction of this spatial information improves the overall classification accuracy. In the case of the Flevoland dataset, containing urban and agricultural areas, the overall accuracy increased from 95.7%, obtained with traditional methods, to 96.1%, whereas in the case of the San Francisco dataset, containing mainly urban areas, the overall accuracy increased from 92.6 to 94.4%. The second paper, authored by H. Zakeri, F. Yamazaki, and W. Liu, proposes the study of the city of Tehran, basically containing urban, bare, and semi-arid areas. The authors aim to classify this urban environment, whose population has increased dramatically, raising from 6 million inhabitants in 1986 to 12 million in 2011. In this case, the authors propose a Support Vector Machine classification scheme based on the use of polarimetric as well as texture information [2]. As in the previous paper, it is demonstrated that the use of spatial information, together with polarimetric information, helps to increase the overall classification accuracy. The third paper, authored by F. Gao, T. Huang, J. Wang, J. Sun, A. Hussain, and E. Yang, also addressed the problem of polarimetric SAR data classification considering spatial information. In this case, the authors propose a dual branch deep convolutional neural network, in which one of the branches considers the classification of the polarimetric information while the second considers the use of the spatial information, also derived from a combination of the original polarimetric images [3]. In this case, the authors also demonstrated that the use of spatial information, in combination with polarimetric information, helps to improve the overall accuracy of the classification approach. As it can be deduced from these interesting contributions, the classification of polarimetric SAR data improves classical classification approaches not based on polarimetric diversity. Nevertheless, the combination of this polarimetric information, together with spatial attributes, seems to offer clear improvement in classification accuracy, as demonstrated by all the authors.

A total of five papers were focused on the applications of SAR polarimetry; four papers were dedicated to land and vegetation, and one was dedicated to ocean. Two papers were devoted to studies on rice, as it is the main staple crop in the world. In the first one [4], Y. Izumi, S. Demirci, M. Z. bin Baharuddin, T. Watanabe, and J. T. Sri Sumantyo analyzed the temporal variations of polarimetric observables derived from full-circular and dual-circular polarimetric data acquired along the rice cultivation cycle with a ground-based sensor, and assessed several variables with regard to their effectiveness in phenology retrieval. O. Yuzugullu, E. Erten, and I. Hajnsek developed in a study [5] on the estimation of biophysical variables in rice fields by employing X-band copolar data and electromagnetic modeling of the scene. Also related to vegetation, but with a broader environmental scope, the paper coauthored by T. Ullmann, S. N. Banks, A. Schmitt, and T. Jagdhuber [6] describes the response of a large number of polarimetric observables, obtained at L-, C-, and X-bands in a tundra scene located in Canada. The use of shorter wavelength imagery (X and C) was beneficial for the characterization of wetland and tundra vegetation, while L-band data highlighted differences between the bare ground classes better. H. Omar, M. A. Misman, and A. R. Kassim [7] addressed the estimation of aboveground biomass in tropical forests by combining dual-pol data from two different sensors, Sentinel-1A and ALOS PALSAR-2, at two different frequency bands. As for the ocean application of polarimetry, Y. Zhang, Y. Li, X. S. Liang, and J. Tsou contributed in their study [8] with a comparison of quad-, compact-, and dual-polarimetry for oil spill classification, in which a new set of input features was proposed and tested.

Finally, one paper was published on polarimetric SAR interferometry, coauthored by D. Lin, J. Zhu, H. Fu, Q. Xie, and B. Zhang [9], in which a truncated singular value decomposition (TSVD)-based method is proposed for forest height inversion from single-baseline PolInSAR data. With such an approach, the assumption of null ground-to-volume ratio in one of the observed channels, common in most PolInSAR algorithms, is avoided.

## 3. Future Trends in Polarimetric SAR

Although this special issue has been closed, more advances in the processing and the exploitation of polarimetric SAR data are expected in the near future. Its usage in conjunction with other data sources (i.e., data fusion) and in a multi-temporal framework (i.e., time series) seems to be the most promising scenario for most final applications. Nonetheless, basic research and theoretical developments are still required to fully quantify the potentials of this remote sensing technology.

**Acknowledgments:** This issue would not be possible without the interesting contributions of many authors, hardworking and professional reviewers, and the dedicated editorial team of *Applied Sciences*. Congratulations to all authors—no matter what the final decisions of the submitted manuscripts were, the feedback, comments, and suggestions from the reviewers and editors helped the authors to improve their papers. We would like to take this opportunity to record our sincere gratefulness to all reviewers.

**Conflicts of Interest:** The authors declare no conflict of interest.

*Appl. Sci.* **2017**, *7*, 768

## References

1. Wang, X.; Cao, Z.; Ding, Y.; Feng, J. Composite Kernel Method for PolSAR Image Classification Based on Polarimetric-Spatial Information. *Appl. Sci.* **2017**, *7*, 612. [CrossRef]
2. Zakeri, H.; Yamazaki, F.; Liu, W. Texture Analysis and Land Cover Classification of Tehran Using Polarimetric Synthetic Aperture Radar Imagery. *Appl. Sci.* **2017**, *7*, 452. [CrossRef]
3. Gao, F.; Huang, T.; Wang, J.; Sun, J.; Hussain, A.; Yang, E. Dual-Branch Deep Convolution Neural Network for Polarimetric SAR Image Classification. *Appl. Sci.* **2017**, *7*, 447. [CrossRef]
4. Izumi, Y.; Demirci, S.; bin Baharuddin, M.Z.; Watanabe, T.; Sri Sumantyo, J.T. Analysis of Dual- and Full-Circular Polarimetric SAR Modes for Rice Phenology Monitoring: An Experimental Investigation through Ground-Based Measurements. *Appl. Sci.* **2017**, *7*, 368. [CrossRef]
5. Yuzugullu, O.; Erten, E.; Hajnsek, I. A Multi-Year Study on Rice Morphological Parameter Estimation with X-Band Polsar Data. *Appl. Sci.* **2017**, *7*, 602. [CrossRef]
6. Ullmann, T.; Banks, S.N.; Schmitt, A.; Jagdhuber, T. Scattering Characteristics of X-, C- and L-Band PolSAR Data Examined for the Tundra Environment of the Tuktoyaktuk Peninsula, Canada. *Appl. Sci.* **2017**, *7*, 595. [CrossRef]
7. Omar, H.; Misman, M.A.; Kassim, A.R. Synergetic of PALSAR-2 and Sentinel-1A SAR Polarimetry for Retrieving Aboveground Biomass in Dipterocarp Forest of Malaysia. *Appl. Sci.* **2017**, *7*, 675. [CrossRef]
8. Zhang, Y.; Li, Y.; Liang, X.S.; Tsou, J. Comparison of Oil Spill Classifications Using Fully and Compact Polarimetric SAR Images. *Appl. Sci.* **2017**, *7*, 193. [CrossRef]
9. Lin, D.; Zhu, J.; Fu, H.; Xie, Q.; Zhang, B. A TSVD-Based Method for Forest Height Inversion from Single-Baseline PolInSAR Data. *Appl. Sci.* **2017**, *7*, 435.

*applied*
*sciences*

MDPI

*Article*

# Composite Kernel Method for PolSAR Image Classification Based on Polarimetric-Spatial Information

Xianyuan Wang [1], Zongjie Cao [1,*], Yao Ding [1] and Jilan Feng [2]

[1] Center for Information Geoscience, University of Electronic Science and Technology of China, Chengdu 611731, China; xywang0307@std.uestc.edu.cn (X.W.); dyao1027@outlook.com (Y.D.)
[2] Huiding Technology Co. Ltd., Chengdu 611731, China; jlfeng911@gmail.com
* Correspondence: zjcao@uestc.edu.cn; Tel.: +86-028-6183-0379

Academic Editors: Carlos López-Martínez and Juan M. Lopez-Sanchez
Received: 14 April 2017; Accepted: 9 June 2017; Published: 13 June 2017

**Abstract:** The composite kernel feature fusion proposed in this paper attempts to solve the problem of classifying polarimetric synthetic aperture radar (PolSAR) images. Here, PolSAR images take into account both polarimetric and spatial information. Various polarimetric signatures are collected to form the polarimetric feature space, and the morphological profile (MP) is used for capturing spatial information and constructing the spatial feature space. The main idea is that the composite kernel method encodes diverse information within a new kernel matrix and tunes the contribution of different types of features. A support vector machine (SVM) is used as the classifier for PolSAR images. The proposed approach is tested on a Flevoland PolSAR data set and a San Francisco Bay data set, which are in fine quad-pol mode. For the Flevoland PolSAR data set, the overall accuracy and kappa coefficient of the proposed method, compared with the traditional method, increased from 95.7% to 96.1% and from 0.920 to 0.942, respectively. For the San Francisco Bay data set, the overall accuracy and kappa coefficient of the proposed method increased from 92.6% to 94.4% and from 0.879 to 0.909, respectively. Experimental results verify the benefits of using both polarimetric and spatial information via composite kernel feature fusion for the classification of PolSAR images.

**Keywords:** PolSAR; image classification; composite kernel; polarimetric features; spatial features; feature fusion

## 1. Introduction

Polarimetric synthetic aperture radar (PolSAR) has become an important remote sensing tool. Besides the advantage of operating in all times and under all weather conditions, it also provides richer ground information than single-polarization SAR [1]. Along with the development of imaging techniques and the enhancing availability of PolSAR data, effective PolSAR image interpretation techniques are an urgent requirement. Land-cover classification is one of the most important tasks of PolSAR image interpretation [2]. Up to now, different PolSAR image classification approaches have been proposed, either based on statistical properties of PolSAR data [3,4] or based on scattering mechanism identification [5,6].

PolSAR images not only include polarimetric information but also include spatial information. Polarimetric or spatial information cannot describe PolSAR image comprehensively, causing information (polarimetric or spatial) loss. Luckily, inspired by the complementarity between spatial and spectral features producing significant improvements in optical image classification [7], in this paper, the main characteristic of the proposed approach is that it takes advantage of both polarimetric and spatial information for classification.

In order to utilize comprehensive polarimetric information, various polarimetric signatures obtained by target decomposition (TD) methods and algebra operations (AO) can be combined together to form a high-dimensional feature space [8–10]. Moreover, so as to take advantage of spatial information, different strategies can be considered, such as using the Markov random field (MRF) model [11,12], classifying with over-segmentation patches [13,14], and exploiting features that contain spatial information [15–17]. Concerning the last strategy, which will be adopted in this paper, one recently emerged method is based on morphological filters [18].

The main problem here is how to make comprehensive use of the two types of features. In this paper, this problem is solved based on the theory of feature fusion via composite kernels [19]. Compared with the traditional feature fusion method using vector stacking [20], composite kernels can tune the contribution of two types of features, exploiting inner properties more sufficiently, regardless of the weight between different features, leading the fusion more naturally.

Our experiments are conducted with real PolSAR data sets: a Flevoland data set and a San Francisco Bay data set. In order to assess the classification performance of the proposed method, user accuracy, overall accuracy, and kappa coefficient are determined. Experimental results demonstrate that the proposed approach can more efficiently exploit both the polarimetric and the spatial information contained in PolSAR images compared with the traditional method of feature fusion. Classification performance is thus improved.

The reminder of the paper is structured as follows. Section 2 introduces the related work of the entire classification, which is made up of three main components: polarimetric features, spatial features, and stacked feature fusion. Section 3 presents a novel feature fusion for the classification of PolSAR images based on composite kernels. In Section 4, the two experiments and data sets are described, and the results and discussion are presented. Finally, Section 5 concludes and discusses potential future studies.

## 2. Related Work

In this section, polarimetric and spatial feature sets are introduced, and the scheme of the construction of those two spaces is given. Then, the traditional method for fusing different feature is described.

### 2.1. Polarimetric and Spatial Features of PolSAR Image

For PolSAR images, there are several representations of the collected data [2]. For original single-look complex (SLC) PolSAR images, at each pixel, the data is stored in the scattering matrix $\mathbf{S} = [S_{hh}, S_{hv}; S_{vh}, S_{vv}]$. According to the reciprocity theorem, we have $S_{hv} = S_{vh}$. Therefore, we can transform the scattering matrix into a vector form $\mathbf{k}_L = [S_{hh} \ \sqrt{2}S_{hv} \ S_{vv}]^T$ using linear basis, where $T$ denotes transpose of a vector. For both training data and test data, the covariance matrix can be further derived as $\mathbf{C} = \langle \mathbf{k}_L \mathbf{k}_L^H \rangle$, where $\langle \cdot \rangle$ denotes the operation of multi-looking, $\mathbf{k}_L^H$ denotes conjugate transpose of $\mathbf{k}_L$. Similarly, another vector form of the SLC PolSAR data can be obtained as $\mathbf{k}_P = [S_{hh} + S_{vv} \ S_{vv} - S_{hh} \ 2S_{hv}]^T / \sqrt{2}$ by using the Pauli basis, and the multi-looking coherency matrix can be derived as $\mathbf{T} = \langle \mathbf{k}_P \mathbf{k}_P^H \rangle$ [2].

In the proposed approach, two feature spaces are constructed in parallel to account for different information contained in the PolSAR images, as shown in Figure 1. For one thing, the polarimetric features describe the pixel-wise scattering mechanisms that are related to the dielectric properties of ground material. For another, the spatial features describe the relationship between neighboring pixels that are related to the ground object structures. Therefore, those two sets of features contain complementary information.

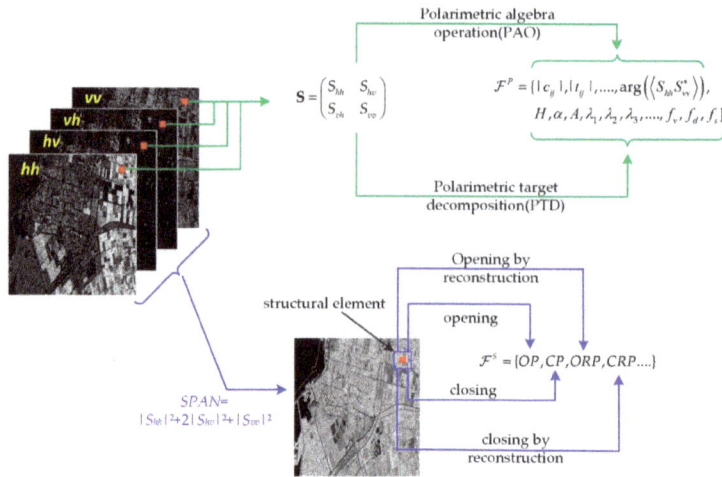

**Figure 1.** The scheme for polarimetric and spatial feature space construction in polarimetric synthetic aperture radar (PolSAR) images. The polarimetric features are concerned with the ground scattering property at a single pixel, while the spatial features are concerned with the relationship between neighboring pixels by using structural elements with different shapes and sizes.

### 2.1.1. Polarimetric Features

The polarimetric feature vector is constructed by collecting various polarimetric signatures obtained by polarimetric algebra operations (PAO) and polarimetric target decomposition (PTD) methods. Here, PAO refers to operations that compute polarimetric signatures with simple mathematical transforms such as summation, difference, and ratio. The derived polarimetric signatures include backscattering intensities, intensity ratios, phase differences, etc. [21]. PTD refers to methods that compute polarimetric signatures with the tool of matrix decompositions [6]. In past decades, many PTD methods have been proposed. According to the data form they deal with, PTD methods can be generally divided into two classes: coherent target decomposition (CTD) methods and incoherent target decomposition (ICTD) methods [1]. While CTD methods deal with the scattering matrix S, ICTD methods deal with the covariance matrix C or the coherency matrix T. CTD methods include Pauli decomposition [6] and Krogager decomposition [22]. Typical ICTD methods include Cloude–Pottier decomposition [23], Yamaguchi decomposition [24], and Freeman–Durden decomposition [25]. Scattering mechanisms related to the dielectric properties of ground material. Different scattering mechanisms were interpreted by the value of parameters in each PTD method. Different PTD methods try to interpret the PolSAR data from different perspectives. Nevertheless, there is no single PTD method that outperforms the others in all cases when used for applications such as classification. To utilize comprehensive polarimetric information for classification, it is advisable to construct a high-dimensional feature vector by collecting various polarimetric signatures. With proper advanced machine learning methods, the discriminative information contained in the high-dimensional feature vectors can be exploited. In this paper, the employed polarimetric signatures are summarized in Table 1. The selection of those polarimetric signatures is based on a survey of several works that make use of multiple polarimetric signatures [8–10] and the literature of [1,2]. Note that we only consider polarimetric signatures extracted from the covariance/coherency matrix because, for multi-looking PolSAR data, the scattering matrix S may not be available.

**Table 1.** Summarization of polarimetric signatures considered in this paper.

| Polarimetric Signatures | Expression |
|---|---|
| Amplitude of upper triangle matrix elements of C | $\left|c_{ij}\right|, 1 \le i \le 3, i \le j \le 3$ |
| Amplitude of upper triangle matrix elements of T | $\left|t_{ij}\right|, 1 \le i \le 3, i \le j \le 3$ |
| Ratio between HV and HH backscattering coefficient | $\langle S_{hv}S_{hv}^*\rangle / \langle S_{hh}S_{hh}^*\rangle$ |
| Ratio between VV and HH backscattering coefficient | $\langle S_{vv}S_{vv}^*\rangle / \langle S_{hh}S_{hh}^*\rangle$ |
| Ratio between HV and VV backscattering coefficient | $\langle S_{hv}S_{hv}^*\rangle / \langle S_{vv}S_{vv}^*\rangle$ |
| Depolarization ratio | $\langle S_{hv}S_{hv}^*\rangle / (\langle S_{hh}S_{hh}^*\rangle + \langle S_{vv}S_{vv}^*\rangle)$ |
| Phase difference HH-VV | $\arg(\langle S_{hh}S_{vv}^*\rangle)$ |
| Entropy, alpha angle, anisotropy and eigenvalues in Cloude Decomposition | $H, \alpha, A$ <br> $\lambda_1, \lambda_2, \lambda_3$ |
| nine parameters of the Huynen Decomposition | $A_0, B_0 + B, B_0 - B,$ <br> $C, D, E, F, G, H$ |
| Power of surface, double-bounce, volume and helix scatter components in Yamaguchi Decomposition | $P_s, P_d, P_v, P_h$ |
| Coefficient for the volume, double bounce and surface components in Van Zyl Decomposition | $f_v, f_d, f_s$ |

2.1.2. Spatial Features

The spatial information in the PolSAR image is captured by using the morphological transformation-based methods, which have been proved to be powerful tool for analyzing remote sensing images [26,27]. Recently, morphological profiles (MPs) has attracted the attention of researchers in image processing due to its excellent ability to describe spatial information. However, little attention has been paid to the use of MPs.

To construct the spatial feature space $\mathcal{F}^S$, a morphological profile (MP) is built for a remote sensing image by using a combination of several morphological operations and a set of structural elements (SEs). Some commonly used morphological operations include erosion, dilation, opening, closing, opening by reconstruction, and closing by reconstruction. Erosion can remove burrs and bumps in images. Dilation can enlarge profiles, filling holes whose size is lower than SEs. The process of opening entails erosion first and dilation second. Closing, on the contrary, entails dilation first and erosion second. Opening and closing can both smoothen an image. Detailed information about morphological operations can be seen in [28]. A summary of morphological operations considered in this paper can be seen in Table 2. Moreover, SEs can have different shapes and sizes, and these different structures present can be captured in images. By combining different morphological operations and SEs in the process of the input image, different structures are emphasized in the resulting images. Since structures essentially reflect the spatial relationship between pixels, spatial information is captured in this process.

To build the MP for a PolSAR image, we apply morphological transformations to the *SPAN* image, which is obtained by [2]

$$SPAN = |s_{hh}|2 + 2|s_{hv}|2 + |s_{vv}|^2 \tag{1}$$

If only multi-looking PolSAR data is available, we generate the base image by computing the trace of the coherence matrix **T** or covariance matrix **C**, which is $\langle|s_{hh}|\rangle^2 + 2\langle|s_{hv}|\rangle^2 + \langle|s_{vv}|\rangle^2$. The *SPAN* image is the non-coherent superposition of three polarimetric channel images and can be considered as the total backscattering power. Compared to each single-channel image, the speckle noise in the *SPAN* image is effectively reduced, which helps to make the actual structure of ground objects more clear. Therefore, the *SPAN* image is more suitable to be taken as the base image to build MP than single-channel images. There is another reason that we do not apply morphological transformations to each channel image. It is shown that, for a multi-channel image, applying morphological transformations to a signal channel image separately may cause loss or corruption of information [29]. Therefore, we use the *SPAN* image to build the MP for PolSAR image classification.

**Table 2.** Summarization of morphological operations.

| Morphological Operations | Expression |
|---|---|
| Erosion | $\varepsilon(I,b)(s,t) = \min\{I(s+x,t+y) - b(x,y)|(s+x,t+y) \in D_I; (x,y) \in D_b\}$ |
| Dilation | $\delta(I,b)(s,t) = \min\{I(s-x,t-y) + b(x,y)|(s-x,t-y) \in D_I; (x,y) \in D_b\}$ |
| Opening | $\gamma_{b_{ero},b_{dil}}(I) = \delta_{b_{dil}}(\varepsilon_{b_{ero}}(I))$ |
| Closing | $\delta_{b_{ero},b_{dil}}(I) = \varepsilon_{b_{dil}}(\delta_{b_{ero}}(I))$ |
| Opening by reconstruction | $\gamma^{rec}_{b_{dil},bc}(I) = \gamma_{bc,J}(I), \, J = \varepsilon_{b_{ero}}(I)$ |
| Closing by reconstruction | $\phi^{rec}_{b_{dil},bc}(I) = \phi_{bc,g}(I), \, g = \delta_{b_{dil}}(I)$ |

$\varepsilon$ and $\delta$ represent erosion and dilation operation, respectively. $b_{dil}$ and $b_{ero}$ represent SE of dilation and erosion. $I$ represent the *SPAN* image.

### 2.2. Stacked Feature Fusion

In this section, the polarimetric and spatial features are used for making feature fusion based on the traditional method for classification of PolSAR images. Figure 2 shows the processing flow of PolSAR image classification with a feature fusion scheme.

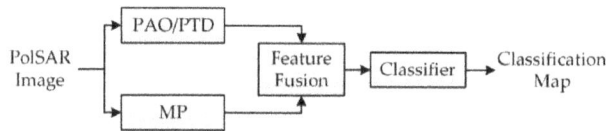

**Figure 2.** Polarimetric-spatial classification of PolSAR images by feature fusion.

In PolSAR image classification, the most commonly adopted approach is to exploit the polarimetric information. However, performance can be improved by concerning both polarimetric and spatial information in the classifier. Now, the problem is how to combine the features to decide the class that a pixel belongs to. Traditionally, the method is based on a "stacked" approach, in which feature the vector is built from the concatenation of polarimetric and spatial features [20,30].

Denote $\mathcal{F}^P$ and $\mathcal{F}^S$ as the polarimetric and spatial feature space, respectively. At a pixel $X_i$, we have two feature vectors $X_i^P \in \mathcal{F}^P$ and $X_i^S \in \mathcal{F}^S$. Two feature vectors $X^P$ and $X^S$ are concatenated to form a new single vector,

$$X^C = [X^P, X^S].\qquad(2)$$

$X^C$ is used for the following SVM-based classification. The advantage of vector stacking based approach is simple to be implemented for different feature vectors. The "stacked" method needs only to concatenate to form a new vector. Unfortunately, from Equation (2), $X^C$ does not include an explicit cross relation between $X^P$ and $X^S$. Therefore, the vector stacking method may lead to a bad performance because the weight and relationship between different features are not considered in this method. When a vector stacking scheme is used, the two feature vectors should be carefully normalized so that their contribution is not affected by the difference in scale.

### 3. Composite Kernels for SVM

In this section, composite kernel feature fusion is proposed to combine different features for PolSAR image classification. SVM and kernels are also briefly described.

## 3.1. SVM and Kernel

Based on the theory of structural risk minimization [31], a support vector machine (SVM) attempts to find a hyper-plane that separates two classes while maximizing the margin between two classes. Nowadays, SVM has become one of the most popular tools for solving pattern recognition problems [32–34]. In remote sensing, SVM is of particular interest to researchers due to its ability to deal with high-dimensional features with a relatively small number of training samples and to handle nonlinear separable issues by using kernel tricks [35,36].

Given a training feature set $\mathcal{T} = \{(X_1, l_1), \dots, (X_N, l_N)\}$, where $X_i$ is the $i$-th feature sample and $l_i$ is the corresponding label (for two class cases we have $l_i \in \{-1, 1\}$), the SVM finds the decision function by solving the following optimization problem:

$$\max_{\alpha} \left\{ \sum_{i=1}^{N} \alpha_i - \frac{1}{2} \sum_{i=1}^{N} \sum_{j=1}^{N} \alpha_i \alpha_j l_i l_j \langle X_i, X_j \rangle \right\}$$
$$s.t. \ 0 \leq \alpha_i \leq C \text{ and } \sum_{i=1}^{N} \alpha_i l_l = 0 \tag{3}$$

where $\alpha = [\alpha_1, \alpha_2, \cdots, \alpha_N]$ is the vector of Lagrange coefficients, $C$ is a constant used for penalize training errors.

Input Space $\mathcal{F}$ (often linear inseparability) can be mapped into high dimensional Hilbert space $\mathcal{H}$, i.e., let $\Phi(X_i)$ replaces $X_i$, and $\Phi(X_i) \in \mathcal{H}$. It is assumed that $X_i$ are more likely to be linear separable in Hilbert space $\mathcal{H}$. Then, define a kernel function $K(X_i, X_j) = \langle \Phi(X_i), \Phi(X_j) \rangle$. In this way, a linear hyper-plane can be found in high-dimensional Hilbert space. The optimization problem becomes

$$\max_{\alpha} \left\{ \sum_{i=1}^{N} \alpha_i - \frac{1}{2} \sum_{i=1}^{N} \sum_{j=1}^{N} \alpha_i \alpha_j l_i l_j \kappa \langle X_i, X_j \rangle \right\}$$
$$s.t. \ 0 \leq \alpha_i \leq C \text{ and } \sum_{i=1}^{N} \alpha_i l_l = 0 \tag{4}$$

where $\kappa$ is a kernel function. The dual formulation of the SVM problem in Equation (4) is convex, which facilitates the solving process.

To predict the label for a new feature sample $X$, the score function of the SVM is computed with the optimal Lagrange coefficients $\alpha^*$ obtained by solving Equation (5):

$$\phi(X) = \sum_{i=1}^{N} \alpha_i^* l_i \kappa(X_i, X) + b \tag{5}$$

where $b$ is the bias of the decision function. For a two-class problem, the label of the test feature sample $X$ is given by $l(X) = \text{sgn}(\phi(X))$. The two-class SVM can be extended to deal with multiclass classification problems by using a one-versus-one rule or a one-versus-all rule. It is also possible to compute probabilistic outputs from the classification scores for SVM. In this paper, the method of SVM is implemented by using the library LIBSVM [37]. The parameters of SVM are set by a cross-validation step [29].

## 3.2. Composite Kernels

Composite kernels we considered are based on the concept of composite kernels [19]. In SVM, the information contained in the features is encoded within the kernel matrix, whose elements are the value of the kernel function between feature vector pairs.

Some popular kernels are as follows:

- Linear kernel: $\kappa(X_i, X_j) = \langle X_i, X_j \rangle$.
- Polynomial kernel: $\kappa(X_i, X_j) = (\langle X_i, X_j \rangle + 1)^d, d \in^+$.

- Radial basis function: $\kappa\left(X_i, X_j\right) = \exp\left(-\|X_i - X_j\|^2/2\sigma^2\right), \sigma \in R^+$

The parameter of a kernel can be set by a cross-validation step [35].

In mathematics, a real-valued function $\kappa(x, y)$ is said to fulfill Mercer's condition if all square integrate functions $g(x)$ have

$$\iint g(x)\kappa(x, y)g(y)dxdy \geq 0. \tag{6}$$

Based on the Hilbert–Schmidt theory, $\kappa(x, y)$ can be a form of dot production if it fulfills Mercer's condition. The linear kernel, polynomial kernel and radial basis function fulfill Mercer's condition [38].

Some properties of Mercer's kernels are as follows:

Let $\kappa_1$ and $\kappa_2$ are two Mercer's kernels, and $\eta > 0$. Thus,

$$\begin{aligned} \kappa\left(X_i, X_j\right) &= \kappa_1\left(X_i, X_j\right) + \kappa_2\left(X_i, X_j\right) \\ \kappa\left(X_i, X_j\right) &= \eta\kappa_1\left(X_i, X_j\right) \end{aligned} \tag{7}$$

are valid Mercer's kernels.

Therefore, it is possible to encode the information contained in polarimetric and spatial features into two kernel matrixes if we use Mercer's kernel. [39].

A composite kernel called weighted summation kernel that balances the polarimetric and spatial weight can be created as follows:

$$\kappa(X_i, X_j) = \eta\kappa^P(X_i{}^P, X_j{}^P) + (1 - \eta)\kappa^S(X_i{}^S, X_j{}^S) \tag{8}$$

where $X_i, X_j$ are two pixels, $\kappa^P$ is the kernel function for polarimetric features, $\kappa^S$ is the kernel function for spatial features, and $\eta \in [0, 1]$ is tuned in the training process and constitutes a tradeoff between the polarimetric and spatial information to classify a given pixel. The composite kernel allows us to extract some information from the best tuned $\eta$ parameter [30].

Based on practical situation, we can set value of the parameter flexibly. In addition, the composite kernel method can make fusion more naturally and effectively, regardless of the weight between different features, bringing better performance.

Once all kernel functions are evaluated and combined, the obtained kernel matrix is fed into a standard SVM for PolSAR image classification.

## 4. Experiment

To demonstrate the superiority of the composite kernel method, in this section, the composite kernel method is compared to the polarimetric feature method (only considering polarimetric features), the morphological profile method (only considering spatial feature), and the vector stacking method. Firstly, we give the information of the used PolSAR data set. Then, we present the experimental setting. Finally, for each PolSAR image, four approaches are applied and evaluated both qualitatively and quantitatively.

### 4.1. Data Description

The test cases are two PolSAR data sets that were collected by the RadarSat-2 system over the area of Flevoland in Netherlands and the AIRSAR system over the area of San Francisco Bay.

### 4.1.1. Flevoland Data Set

The first data set is comprised of a C-band RadarSat-2 PolSAR image collected in the fine quad-pol mode. To facilitate visual interpretation and evaluation, a nine-look multi-looking processing was performed before used for classification. A subset with a size of $700 \times 780$ pixels was used in our experiment. Figure 3a is a color image obtained by Pauli decomposition, and Figure 3b is the corresponding reference map. A total of four classes are identified: building area, woodland, farmland,

and water area. Note that the reference map is not exhaustive, and pixels that are not assigned to any class are shown in gray. To use the data set for performance evaluation, labeled pixels are randomly split into two sets that are used as the training set and the testing set. To ensure the stability and reliability of the performance evaluation results while keeping the computational burden within a controllable range, 1% of the labeled pixels are selected as training samples and the rest of the labeled pixels are taken as testing samples. Detailed information about the training and testing set is listed in Table 3.

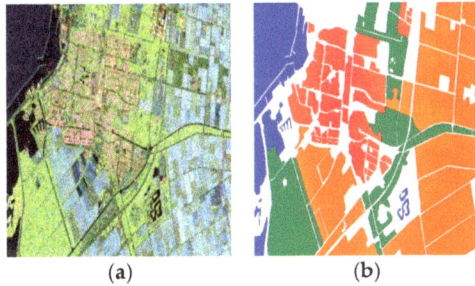

**Figure 3.** Flevoland data set collected by RadarSat-2. (**a**) RGB image obtained by Pauli decomposition. (**b**) Reference map. A total of four classes are identified. Color-coded class label: red—building area, green—woodland, orange—farmland, and blue—water.

**Table 3.** Number of samples of the Flevoland data set used for quantitative evaluation.

| Class | Building | Woodland | Farmland | Water |
|---|---|---|---|---|
| Number of Samples in the Reference Map | 71,331 | 85,539 | 184,920 | 59,504 |
| Number of Training Samples | 713 | 855 | 1849 | 595 |
| Number of Testing Samples | 70,618 | 84,684 | 183,071 | 58,909 |

#### 4.1.2. San Francisco Bay Data Set

The second data set regarding San Francisco Bay is comprised of NASA/JPL AIRSAR L-band PolSAR data. The size of the image is $900 \times 1024$. In Figure 4a, the color image obtained with Pauli decomposition is shown. This image is manually labeled with three classes: city, vegetation, and water. The obtained reference map is shown in Figure 4b, in which different colors stands for different land cover types. Similar to the Flevoland data set, we randomly split pixels into the training set and the testing set. Detailed information about the training and testing set is listed in Table 4.

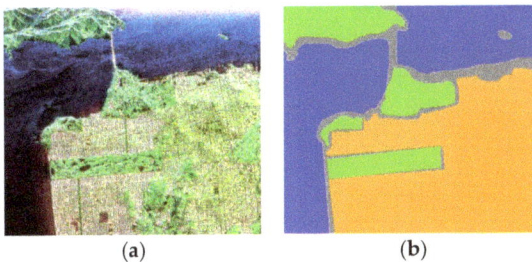

**Figure 4.** San Francisco Bay data set collected by AIRSAR system. (**a**) RGB image obtained by Pauli decomposition. (**b**) Reference map. A total of three classes are identified. Color-coded class label: yellow—city area, blue—water, and green—vegetation.

**Table 4.** Number of samples of the San Francisco Bay data set used for quantitative evaluation.

| Class | City Area | Water | Vegetation |
|---|---|---|---|
| Number of Samples in the Reference Map | 391,407 | 315,320 | 135,508 |
| Number of Training Samples | 3914 | 3153 | 1355 |
| Number of Testing Samples | 387,439 | 312,167 | 134,153 |

*4.2. General Setting*

The polarimetric features considered in this paper are summarized in Table 1. For morphological feature extraction, opening, closing, opening by reconstruction, and closing by reconstruction are considered and are summarized in Table 2. For each of those filters, an SE whose dimensions unit increased from 5 to 19 with a step of 2 pixels is used. In the classification stage, the involved parameters are parameters in the SVM, and the weight parameter $\eta$ that can tune the contribution of two kernels. And the kernel we used is Radial basis function (RBF) which fulfills Mercer's condition.

In the training process, the overall accuracy curve in Figure 5a reaches the top when $\eta = 0.6$. So, we may conclude that set $\eta = 0.6$ in Flevoland data set can make the best performance. Similarly, in Figure 5b, setting $\eta = 0.6$ in San Francisco Bay data set had best outcome. Therefore, we set $\eta = 0.6$ in the experiment for the Flevoland and San Francisco Bay data sets.

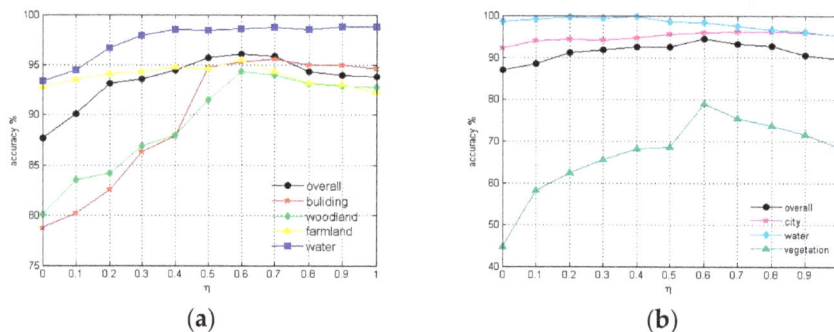

**Figure 5.** Accuracy versus weight $\eta$ for Flevoland data set (**a**) and San Francisco Bay data set (**b**).

*4.3. Results*

4.3.1. Flevoland Data Set

As expected, we can see Figure 6a that, regarding polarimetric features alone, the building area and woodland area in black ellipses show large variation, i.e., pixels of the same land cover type may have very different scattering mechanisms, while pixels of different land cover types may have very similar scattering mechanisms. Moreover, in Figure 6b, morphological features alone may also lead to incorrect classification results, e.g., those two farmland areas in white ellipses that are classified as woodland area. Further, feature fusion via vector stacking also has some flaws; e.g., in Figure 6c, the farmland in yellow ellipses shows obvious variation. However, when combining those two features via composite kernels, the performance of the PolSAR image improves; e.g., in Figure 6d, in black ellipses, building and woodland can be clearly classified, in white ellipses, farmland can be classified, and in yellow ellipses, farmland can also be classified—much better than the vector stacking method. And Figure 6e is the reference map of Flevoland Data Set.

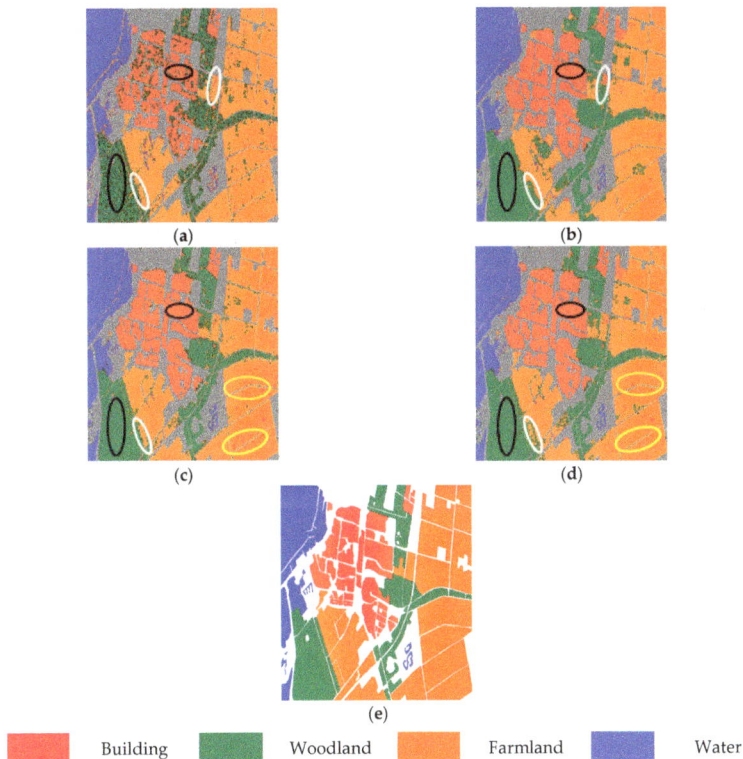

**Figure 6.** Classification maps of Flevoland data with different methods. (**a**) Only polarimetric features. (**b**) Only morphological features. (**c**) Feature fusion via vector stacking. (**d**) Feature fusion via composite kernels. (**e**) Reference map.

From Table 5, we can see that using polarimetric features alone lead to low accuracy, especially in building areas and woodland areas. Morphological features account for the spatial structural information, which helps to produce more accurate classification results. However, this approach has a slightly low accuracy in farmland area because this approach only considers spatial information. The overall accuracy is distinctly boosted when fusing the two types of information together. The kappa coefficient also notably increases to 0.94. Admittedly, the overall accuracy of vector stacking is similar to that of the composite kernel approach, but it is worth noting that, when the vector stacking method is used, different features should be normalized before stacking. Further, the vector stacking approach does not take the inter-relation of different features into consideration.

**Table 5.** Classification accuracy measures on the Flevoland data set. The best performance in each column is shown in bold.

| Method | Building (%) | Woodland (%) | Farmland (%) | Water (%) | Overall Accuracy (%) | Kappa Coefficient |
|---|---|---|---|---|---|---|
| Only POL | 78.8 | 80.1 | 92.8 | 93.4 | 87.7 | 0.842 |
| Only MP | 94.6 | 92.7 | 92.3 | **98.9** | 93.8 | 0.909 |
| Vector Stacking | 95.2 | 93.7 | 96.2 | 97.4 | 95.7 | 0.920 |
| Composite Kernel | **95.6** | **94.3** | **96.2** | 98.8 | **96.1** | **0.942** |

Kappa Coefficient measures the percentage of data values in the main diagonal of the table and then adjusts these values for the amount of agreement that could be expected due to chance alone [40].

4.3.2. San Francisco Bay Data Set

Like the Flevoland data set, firstly, as shown in Figure 7a, the city and water area in red ellipses show large variation. Secondly, in Figure 7b, though morphological features alone show a smoother map, they may also lead to an incorrect classification map, e.g., the two vegetation areas in white ellipses that are classified as water areas. Moreover, feature fusion via vector stacking also has some flaws; e.g., in the upper right (luminous yellow ellipses) of Figure 7c, there are areas and variations that are obviously incorrectly classified. However, when combining those two features via composite kernels, the performance of the PolSAR image improves, and the map of the composite kernel method (in Figure 7d) is smoother and has more correctly classified areas than those of other methods. And Figure 7e is the reference map of San Francisco Bay Data Set.

From Table 6, the overall accuracy of the four methods (only POL, only MP, vector stacking, and composite kernels) is 86.9%, 89.6%, 92.6, and 94.4%, respectively. Compared with the other three methods, the overall accuracy of the new method we proposed is increased by 7%, 4.8%, and 1.8%. The kappa coefficient also exhibits marked growth.

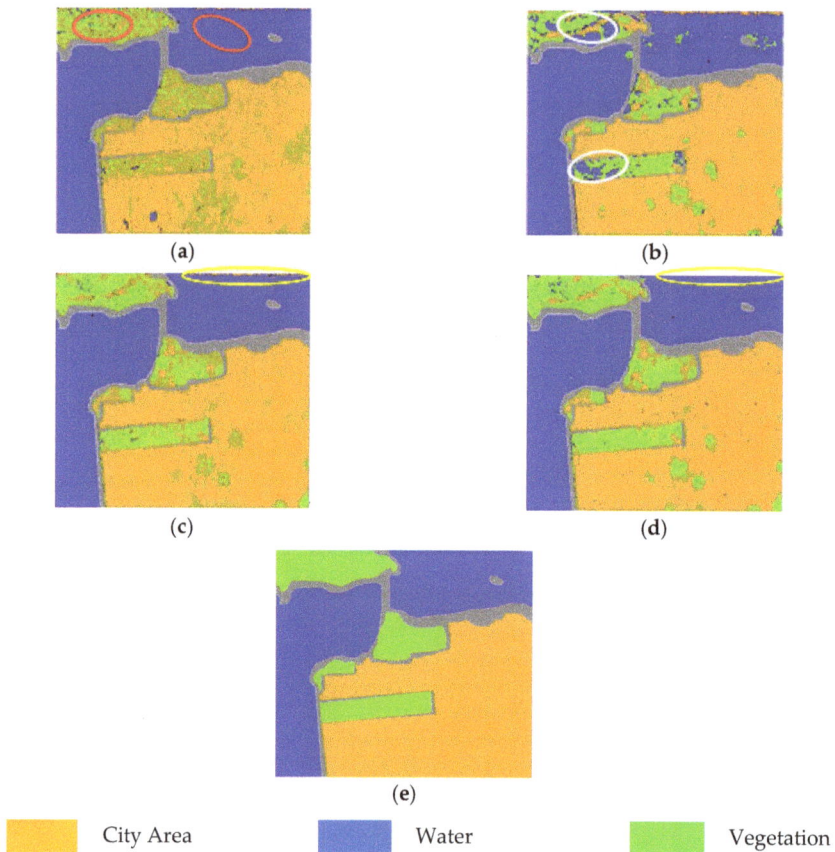

**Figure 7.** Classification maps of the San Francisco Bay data set with different methods. (**a**) Only polarimetric features. (**b**) Only morphological features. (**c**) Feature fusion via vector stacking. (**d**) Feature fusion via composite kernels. (**e**) Reference map.

**Table 6.** Classification accuracy measures on San Francisco Bay data set. The best performances in each column are shown in bold.

| Method | City Area (%) | Water (%) | Vegetation (%) | Overall Accuracy (%) | Kappa Coefficient |
|---|---|---|---|---|---|
| Only POL | 92.2 | 98.5 | 44.7 | 86.9 | 0.783 |
| Only MP | 95.1 | 95.2 | 60.7 | 89.6 | 0.830 |
| Vector Stacking | **96.0** | 98.5 | 68.5 | 92.6 | 0.879 |
| Composite Kernel | **96.0** | **99.7** | **78.8** | **94.4** | **0.909** |

Our results confirm the validation of the composite kernel method. The performance of classification can be boosted when we fuse two types of information. The composite kernel method can tune the contribution of different content, yielding better accuracy than the "stacked" method. However, we have not considered other possible kernel distances. The classification performance may be improved if we take other kernels into account.

## 5. Conclusions and Future Work

In this paper, a feature fusion approach that exploits both polarimetric and spatial information of PolSAR images is proposed. The polarimetric information is captured by collecting polarimetric signatures. The spatial information is captured by using the morphological transformation. Inspired by the complementarity between spatial and spectral features producing significant improvements in optical image classification, performance can be improved by fusing polarimetric and spatial information in the classifier. Traditionally, the method is based on a "stacked" approach, in which feature vectors are built from the concatenation of polarimetric and spatial features. In this paper, we propose a new method called the "composite kernel" method, which tunes the contribution of different type of features. Compared with one signal feature classification and traditional feature fusion classification via vector stacking, the composite kernel method possesses several excellent properties, making fusion more effective and leading to better performance. The proposed approach has been tested on the Flevoland data set and the San Francisco Bay data set. The obtained results confirm the benefit of combing different types of information via composite kernels for PolSAR image classification.

In the future, we will continue studying PolSAR image classification with polarimetric and spatial features. Other possible kernel distances could be investigated via, for example, the spectral angle mapper. More advanced feature combination methods that solve the problems of feature extraction and combination simultaneously could also be researched. One such framework is multiple kernel learning, which could be utilized as a platform for developing effective classification approaches.

**Acknowledgments:** This study was supported by the Fundamental Research Funds for the Central Universities A03013023601005 and the National Nature Science Foundation of China under Grant 61271287 and Grant U1433113.

**Author Contributions:** Zongjie Cao led the study. Xianyuan Wang designed the experimental and wrote the paper. Yao Ding arranged the data sets. Jilan Feng analyzed the data and checked the paper.

**Conflicts of Interest:** The authors declare no conflict of interest.

## References

1. Touzi, R.; Boerner, W.M.; Lee, J.S.; Lueneburg, E. A review of polarimetry in the context of synthetic aperture radar: Concepts and information extraction. *Can. J. Remote Sens.* **2004**, *30*, 380–407. [CrossRef]
2. Lee, J.-S.; Pottier, E. *Polarimetric Radar Imaging: From Basics to Applications*; CRC Press: Boca Raton, FL, USA, 2009.
3. Doulgeris, A.P.; Anfinsen, S.N.; Eltoft, T. Automated non-Gaussian clustering of polarimetric synthetic aperture radar images. *IEEE Trans. Geosci. Remote Sens.* **2001**, *49*, 3665–3676. [CrossRef]
4. Liu, H.-Y.; Wang, S.; Wang, R.-F. A framework for classification of urban areas using polarimetric SAR images integrating color features and statistical model. *J. Infrared Millim. Waves* **2016**, *35*, 398–406.

5. Fan, K.T.; Chen, Y.S.; Lin, C.W. Identification of rice paddy fields from multitemporal polarimetric SAR images by scattering matrix decomposition. In Proceedings of the IEEE International Symposium on Geoscience and Remote Sensing IGARSS, Milan, Italy, 26–31 July 2015; pp. 3199–3202.
6. Cloude, S.R.; Pottier, E. A review of target decomposition theorems in radar polarimetry. *IEEE Trans. Geosci. Remote Sens.* **1996**, *34*, 498–518. [CrossRef]
7. Yuan, H.; Tang, Y.Y. Spectral-Spatial Shared Linear Regression for Hyperspectral Image Classification. *IEEE Trans. Cybern.* **2017**, *47*, 934–945. [CrossRef] [PubMed]
8. Zou, T.; Yang, W.; Dai, D.; Sun, H. Polarimetric SAR Image Classification Using Multifeatures Combination and Extremely Randomized Clustering Forests. *EURASIP J. Adv. Signal Process.* **2010**, *2010*, 4. [CrossRef]
9. Tu, S.T.; Chen, J.Y.; Yang, W.; Sun, H. Laplacian eigenmaps based polarimetric dimensionality reduction for SAR image classification. *IEEE Trans. Geosci. Remote Sens.* **2012**, *50*, 170–179. [CrossRef]
10. Shi, L.; Zhang, L.; Yang, J. Supervised graph embedding for polarimetric SAR image classification. *IEEE Geosci. Remote Sens. Lett.* **2013**, *10*, 216–220. [CrossRef]
11. Li, J.; Bioucas-Dias, J.M.; Plaza, A. Spectral-spatial hyperspectral image segmentation using subspace multinomial logistic regression and Markov random fields. *IEEE Trans. Geosci. Remote Sens.* **2012**, *50*, 809–823. [CrossRef]
12. Moser, G.; Serpico, S.; Benediktsson, J.A. Land-cover mapping by Markov modeling of spatial–contextual information in very-high-resolution remote sensing images. *Proc. IEEE* **2013**, *101*, 631–651. [CrossRef]
13. Chen, X.; Fang, T.; Huo, H.; Li, D. Graph-based feature selection for object-oriented classification in VHR airborne imagery. *IEEE Trans. Geosci. Remote Sens.* **2011**, *49*, 353–365. [CrossRef]
14. Liu, B.; Hu, H.; Wang, H.; Wang, K.; Liu, X.; Yu, W. Superpixel-based classification with an adaptive number of classes for polarimetric SAR images. *IEEE Trans. Geosci. Remote Sens.* **2013**, *51*, 907–924. [CrossRef]
15. Dai, D.; Yang, W.; Sun, H. Multilevel local pattern histogram for SAR image classification. *IEEE Geosci. Remote Sens. Lett.* **2011**, *8*, 225–229. [CrossRef]
16. Shen, L.; Zhu, Z.; Jia, S.; Zhu, J.; Sun, Y. Discriminative Gabor feature selection for hyperspectral image classification. *IEEE Geosci. Remote Sens. Lett.* **2013**, *10*, 29–34. [CrossRef]
17. Huang, X.; Zhang, L. An SVM ensemble approach combining spectral, structural, and semantic features for the classification of high-resolution remotely sensed imagery. *IEEE Trans. Geosci. Remote Sens.* **2013**, *51*, 257–272. [CrossRef]
18. Soille, P. *Morphological Image Analysis*; Springer: Berlin, Germany, 2004.
19. Devis, T.; Frederic, R.; Alexei, P.; Camps-Valls, G. Composite kernels for Urban-image classification. *IEEE Geosci. Remote Sens. Lett.* **2010**, *7*, 88–92.
20. Du, P.; Samat, A.; Waskec, B.; Liu, S.; Li, Z. Random Forest and Rotation Forest for fully polarized SAR image classification using polarimetric and spatial features. *ISPRS J. Photogramm. Remote Sens.* **2015**, *105*, 38–53. [CrossRef]
21. Molinier, M.; Laaksonen, J.; Rauste, Y.; Häme, T. Detecting changes in polarimetric SAR data with content-based image retrieval. In Proceedings of the IEEE International Geoscience and Remote Sensing Symposium, Barcelona, Spain, 23–28 July 2007; pp. 2390–2393.
22. Krogager, E. New decomposition of the radar target scattering matrix. *Electron. Lett.* **1990**, *26*, 1525–1527. [CrossRef]
23. Cloude, S.R.; Pottier, E. An entropy based classification scheme for land applications of polarimetric SAR. *IEEE Trans. Geosci. Remote Sens.* **1997**, *35*, 68–78. [CrossRef]
24. Yamaguchi, Y.; Moriyama, T.; Ishido, M.; Yamada, H. Four component scattering model for polarimetric SAR image decomposition. *IEEE Trans. Geosci. Remote Sens.* **2005**, *43*, 1699–1706. [CrossRef]
25. Freeman, A.; Durden, S.L. A three-component scattering model for polarimetric SAR data. *IEEE Trans. Geosci. Remote Sens.* **1998**, *36*, 963–973. [CrossRef]
26. Soille, P.; Pesaresi, M. Advances in mathematical morphology applied to geoscience and remote sensing. *IEEE Trans. Geosci. Remote Sens.* **2002**, *40*, 2042–2055. [CrossRef]
27. Fauvel, M.; Benediktsson, J.A.; Chanussot, J.; Sveinsson, J.R. Spectral and spatial classification of hyperspectral data using SVMs and morphological profiles. *IEEE Trans. Geosci. Remote Sens.* **2008**, *46*, 3804–3814. [CrossRef]
28. Zhao, Y.-Q.; Liu, J.-X.; Liu, J. Medical image segmentation based on morphological reconstruction operation. *Comput. Eng. Appl.* **2007**, *43*, 228–240.

*Appl. Sci.* **2017**, *7*, 612

29. Plaza, A.; Martinez, P.; Plaza, J.; Perez, R. Dimensionality reduction and classification of hyperspectral image data using sequences of extended morphological transformations. *IEEE Trans. Geosci. Remote Sens.* **2005**, *43*, 466–479. [CrossRef]

30. Camps-Valls, G.; Gomez-Chova, L.; Muñoz-Mari, J.; Vila-Frances, J.; Calpe-Maravilla, J. Composite kernels for hyperspectral image classification. *IEEE Geosci. Remote Sens. Lett.* **2006**, *3*, 93–97. [CrossRef]

31. Vapnik, V.N. *The Nature of Statistical Learning Theory*, 2nd ed.; Springer: New York, NY, USA, 2000.

32. Burges, C. A tutorial on support vector machines for pattern recognition. *Data Min. Knowl. Discov.* **1998**, *2*, 121–167. [CrossRef]

33. Gu, B.; Sheng, V.S. A Robust Regularization Path Algorithm for ν-Support Vector Classification. *IEEE Trans. Neural Netw. Learn. Syst.* **2016**, *28*, 1241–1248. [CrossRef] [PubMed]

34. Gu, B.; Sheng, V.S.; Tay, K.Y.; Romano, W.; Li, S. Incremental Support Vector Learning for Ordinal Regression. *IEEE Trans. Neural Netw. Learn. Syst.* **2015**, *26*, 1403–1416. [CrossRef] [PubMed]

35. Kim, E. Every You Wanted to Know about the Kernel Trick (But Were Too Afraid to Ask). Available online: http://www.eric-kim.net/eric-kim-net/posts/1/kernel_trick.html (accessed on 1 June 2016).

36. Gu, B.; Sheng, V.S.; Wang, Z.; Ho, D.; Osman, S.; Li, S. Incremental learning for ν-Support Vector Regression. *Neural Netw.* **2015**, *67*, 140–150. [CrossRef] [PubMed]

37. LIBSVM Tools. Available online: http://www.csie.ntu.edu.tw/~cjlin/libsvmtools (accessed on 1 June 2016).

38. Mercer's Theorem. Available online: https://en.wikipedia.org/wiki/Mercer%27s_theorem (accessed on 1 June 2016).

39. Tuia, D.; Ratle, F.; Pozdnoukhov, A.; Camps-valls, G. Multisource composite kernels for urban-image classification. *IEEE Geosci. Remote Sens. Lett.* **2010**, *7*, 88–92. [CrossRef]

40. Kappa Coefficients: A Critical Appraisal. Available online: http://john-uebersax.com/stat/kappa.htm (accessed on 1 June 2016).

*applied*
*sciences*

MDPI

*Article*

# Texture Analysis and Land Cover Classification of Tehran Using Polarimetric Synthetic Aperture Radar Imagery

**Homa Zakeri \*, Fumio Yamazaki and Wen Liu**

Department of Urban Environment Systems, Chiba University, Chiba 263-8522, Japan;
fumio.yamazaki@faculty.chiba-u.jp (F.Y.); wen.liu@chiba-u.jp (W.L.)
\* Correspondence: homa.zakeri@alumni.ut.ac.ir; Tel.: +81-43-290-3528

Academic Editors: Carlos López-Martínez and Juan Manuel Lopez-Sanchez
Received: 9 March 2017; Accepted: 21 April 2017; Published: 29 April 2017

**Abstract:** Land cover classification of built-up and bare land areas in arid or semi-arid regions from multi-spectral optical images is not simple, due to the similarity of the spectral characteristics of the ground and building materials. However, synthetic aperture radar (SAR) images could overcome this issue because of the backscattering dependency on the material and the geometry of different surface objects. Therefore, in this paper, dual-polarized data from ALOS-2 PALSAR-2 (HH, HV) and Sentinel-1 C-SAR (VV, VH) were used to classify the land cover of Tehran city, Iran, which has grown rapidly in recent years. In addition, texture analysis was adopted to improve the land cover classification accuracy. In total, eight texture measures were calculated from SAR data. Then, principal component analysis was applied, and the first three components were selected for combination with the backscattering polarized images. Additionally, two supervised classification algorithms, support vector machine and maximum likelihood, were used to detect bare land, vegetation, and three different built-up classes. The results indicate that land cover classification obtained from backscatter values has better performance than that obtained from optical images. Furthermore, the layer stacking of texture features and backscatter values significantly increases the overall accuracy.

**Keywords:** land cover; supervised classification; texture measures; synthetic aperture radar (SAR) imagery; support vector machine; maximum likelihood; Tehran

## 1. Introduction

Land cover monitoring of urban areas provides vital information in several fields, such as environmental science, seismic risk assessment, urban management, and regional planning. For instance, in sustainable development, urban growth assessment plays an essential role in maintaining the balance between the city and the hinterland. Urban expansion results in the change of urban land cover and the expansion of a city's border, which is necessary for accommodating a growing population and providing them with public city services [1]. Furthermore, it is important to continuously update land cover maps at macro- and micro-scales, which helps governments to be prepared for emergency monitoring of cities, especially after natural hazards [2–11].

Remote sensing technologies are fundamental tools used to obtain information from the ground surface to determine land cover classification. Thus, satellite imagery can be used to analyze urban growth and land cover changes [12]. Moreover, urban residential areas with different properties and densities can be identified [13,14]. One of the most advanced technologies is synthetic aperture radar (SAR) sensors, which have several advantages over optical sensors that are used to capture land surface imagery. SAR sensors can extract object characteristics from backscattering echo, independent of weather conditions and time [15–17]. Currently, this technology, with dual- or full-polarization

(HH, HV, VV and VH), is used widely to monitor urban areas and map land cover since different polarizations have different sensitivities and scattering coefficients for the same target [3,18–21].

Furthermore, texture features represent a significant source of information regarding the spatial relation of pixel values. Thus, different features, such as built-up urban areas, soil, rock, and vegetation, can be more accurately characterized. Many texture measures have been developed and properly used in satellite image analyses. Thus, it is generally accepted that the use of textural images improves the accuracy of land cover classification [22–25]. Previous research on texture feature extraction showed that the gray-level co-occurrence matrix (GLCM) is one of the most trustworthy methods for classification [26,27].

Tehran, the capital city of Iran, has been undergoing rapid changes in land cover and land use, similar to many other metropolitan areas in developing countries. The population of Tehran increased from 6,758,845 in 1996 to 12,183,391 in 2011 [28], almost doubling in only 15 years. For that reason, Tehran is considered as one of the fastest growing cities in the world. Therefore, urban area monitoring of Tehran seems necessary. However, because of the similarity of spectral signatures between soil and roof materials in the built-up areas of Tehran, the accuracy of land cover classification using optical images is expected to be not so high. On the other hand, a SAR image analysis using backscattering intensity data has the potential to accurately classify urban areas.

In this study, the capabilities of SAR images to recognize built-up areas from bare land in Tehran city, Iran, are evaluated. For this purpose, ALOS-2 PALSAR-2 (HH, HV) and Sentinel-1 C-SAR (VV, VH) are used for land cover classification. Texture measures are applied to the backscatter values of the L- and C-bands of the mentioned satellite. Then, supervised land cover classification of Tehran is carried out using the backscattering intensity and texture measures selected by a principal components analysis. This study attempts to examine the performance of SAR intensity data for land cover classification in arid and semi-arid regions.

## 2. Study Area and Dataset

### 2.1. Study Area

The study area is located in Tehran, the capital city of Iran, which is a part of the Tehran metropolitan area located at longitude 51°25′17.44″ E and latitude 35°41′39.80″ N, as shown in Figure 1. Tehran is situated in north-central Iran at the foot of the Alborz Mountains, and places on the sloping ground from the mountains in the north and flat areas near the Great Salt Desert in the south. As shown in Table 1, the population of the city slightly increased from 6,058,207 in 1986 to 6,758,845 in 1996, but it rose significantly to 12,183,391 by 2011. Therefore, the city needed more facilities for the residents. Due to this matter, several land covers and land uses emerged or changed into different ones. The area used for assessing the accuracy of the classification results is in the northwest of Tehran, in District 22, located at longitude 51°5′10″–51°20′40″ and latitude 35°32′16″–35°57′19″. It is the district with the greatest development because urban growth in the western regions of the city is necessary to accommodate the population of downtown areas. Tehran is divided into 22 districts, and each district is sponsored by its specific municipality. Moreover, the residential density in this region includes low (100 buildings per hectare), medium (135 buildings per hectare), and high (200 buildings per hectare) densities [28–30]. Considering the variety of built-up areas, densities, vegetation, and bare land in this region it was chosen for accuracy assessment. Besides, there are few green areas in Tehran city, in which one of the largest one is located in this region.

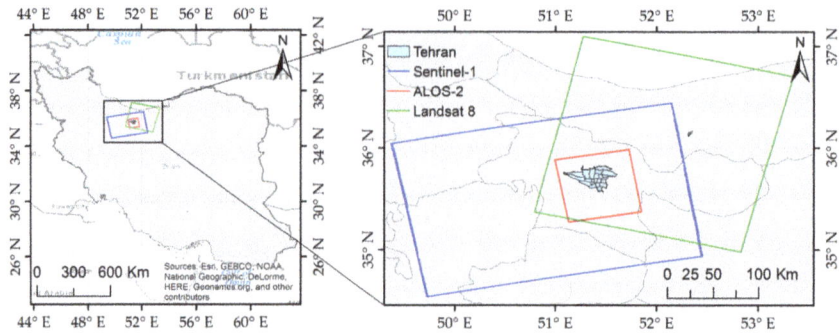

**Figure 1.** Location of Tehran and coverage of the satellite images used in this study, including Sentinel-1 (blue frame) acquired on 26 October 2015, ALOS-2 (red frame) acquired on 14 October 2015 and Landsat 8 (green frame) acquired on 7 May 2015.

**Table 1.** The change of population in Tehran [28].

| Year | 1986 | 1991 | 1996 | 2006 | 2011 |
|------|------|------|------|------|------|
| Population | 6,058,207 | 6,497,238 | 6,758,845 | 7,711,230 | 12,183,391 |

## 2.2. Dataset

The data used in this research were acquired by the Landsat 8, ALOS-2 and Sentinel-1 satellites, which are operated by the National Aeronautics and Space Administration (NASA), the Japan Aerospace Exploration Agency (JAXA) and the European Space Agency (ESA), respectively (Figure 1). Landsat 8 was launched on 11 February 2013 with an Operational Land Imager (OLI) sensor and a Thermal Infrared Sensor (TIRS). The Landsat 8 image was acquired on 7 May 2015 and includes a panchromatic band with 15-m resolution and 11 multi-spectral bands with 30-m resolution. ALOS-2 was launched on 24 May 2014 with an enhanced L-band SAR sensor, PALSAR-2. Its center frequency is 1.27 GHz/23.60 cm in Strip Map mode (SM). Sentinel-1 was launched on 3 April 2014 with the C-SAR sensor in the C-band with a center frequency of 5.40 GHz/5.55 cm in Interferometric Wide Swath mode (IW). Both SAR images covering Tehran were acquired in the ascending path in the right-look direction and by dual-polarization. The ALOS-2 image acquired on 14 October 2015 has HH and HV polarizations, with an incident angle of 40.56° at the center of the image and a spatial resolution of 6.2 m. The Sentinel-1 image captured on 26 October 2015 has VV and VH polarizations with an incident angle of 34.02° at the center of the image and a spatial resolution of 13.9 m. Moreover, the ground swath widths of ALOS-2 and Sentinel-1 are approximately 50–70 km and 250 km, respectively. Both SAR images cover the entire study area in one scene.

## 2.3. Pre-Processing

The SAR images were provided as single-look complex (SLC) data with processing levels of 1.1 for ALOS-2 and 1.0 for Sentinel-1. Both images were represented by the complex I and Q channels to preserve the amplitude and phase information [31,32]. The Sentinel free open source toolboxes were employed for pre-processing. The images were projected on the WGS84 reference ellipsoid. The radiometric calibration of each intensity image was carried out to obtain the backscattering coefficient (sigma-naught, $\sigma^0$) in the ground range with the decibel (dB) unit, represented by the following equation:

$$\sigma^0 = 10.0 \log_{10}\left(k_s \cdot |DN|^2\right) + 10.0 \log_{10}(\sin \theta_{loc}) \tag{1}$$

where *DN* is the digital number of backscattering intensity, $k_s$ is the calibration factor, and $\theta_{loc}$ is the local incidence angle.

After this conversion, different processes were applied to the SAR data. To represent the images as geometrically similar to the real world as possible, geometric terrain correction was applied on the ALOS-2 data using the range Doppler orthorectification method. The Shuttle Radar Topography Mission (SRTM) data were introduced as 3 arc second (approximately 90-m resolution) digital elevation model (DEM). Then, the images were resampled using bilinear interpolation after the terrain correction. As the last step, an adaptive Lee filter with a window size of 3 × 3 [33] was applied to the original polarized SAR images from ALOS-2 to reduce the speckle noise.

Since the IW mode images of Sentinel-1 include three sub-swaths, the Terrain Observation with Progressive Scan SAR (TOPSAR) deburst technique was used to produce a homogenous image for each polarization. Then, the orbit state vectors of Sentinel, precise to the third polynomial degree, were applied to provide accurate information on satellite position and data velocity. Afterwards, the same geometric terrain correction and speckle filtering methods applied to the ALOS-2 images were used. During geometric correction, the pixel size of the Sentinel-1 data was changed to the same pixel size as for the ALOS-2 polarized images (6.25 m). Finally, the two SAR images were registered with the nearest neighbor resampling type and bilinear interpolation method, where the ALOS-2 image was selected as the master.

Additionally, radiometric calibration was applied to the optical image to convert the DN value to the Top of Atmosphere (TOA) reflectance. The image was projected onto the WGS84 reference ellipsoid. Then, the 15-m multi-spectral image was obtained using the pan-sharpening process.

## 3. Methodology

The overall framework is shown in Figure 2. The methodology consists of 5 steps: pre-processing, texture analysis, principal component analysis (PCA), supervised classification, and accuracy assessment. The two sets of polarized SAR images, ALOS-2 and Sentinel-1, were applied in this framework, and their results were evaluated by comparison to truth data. Due to the lack of the real ground truth data in the study area, the Landsat 8 optical image was considered as a base map for preparing the truth data. The truth data contains the polygons of different land covers. All processes in this section were done using Environment for Visualizing Images (ENVI) 5.3.1 software (Exelis Visual Information, Boulder, CO, USA).

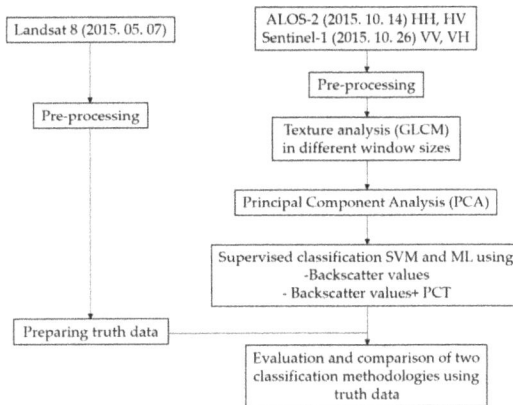

**Figure 2.** Flowchart of land cover classification using the multi-spectral optical image and two synthetic aperture radar (SAR) datasets. GLCM: gray-level co-occurrence matrix; SVM: support vector machine; ML: Maximum Likelihood; PCT: principal components of the textures.

## 3.1. Spatial Texture Analysis

After performing the pre-processing steps, texture measures were applied to each backscattering element (HH, HV, VV and VH). Previous research has shown that texture measures provide vital information from radar imagery [20,34]. Among several statistical texture methods previously proposed, the gray-level co-occurrence matrix (GLCM) is one of the most powerful for land cover monitoring; thus, the GLCM is used in this study. Texture measures represent the spatial distribution of the gray-level value and its frequency relative to another one for a specific displacement at $(x, y)$ and orientation ($0°$, $45°$, $90°$ and $135°$). From a sub-image of a given window size, $I(x, y)$, the GLCM is a matrix $P$ with size $Ng \times Ng$ ($Ng$: the number of gray-levels) whose $P(i, j)$ element ($1 \leq i \leq Ng$; $1 \leq j \leq Ng$) contains the number of times a point with gray-level $g_i$ occurs in a set of positions relative (based on the displacement and the angle mentioned before) to another point with gray-level $g_i$ [22]. The texture measures are calculated from the matrix $P$ as follows:

$$\text{Angular Second Moment (ASM)} = \sum_i \sum_j \{P(i,j)\}^2 \tag{2}$$

$$\text{Contrast} = \sum_{n=0}^{N_g-1} n^2 \left\{ \sum_{\substack{i=1 \\ |i-j|=n}}^{N_g} \sum_{j=1}^{N_g} P(i,j) \right\} \tag{3}$$

$$\text{Correlation} = \frac{\sum_i \sum_j (ij)P(i,j) - \mu_x - \mu_y}{\sigma_x \sigma_y} \tag{4}$$

$$\text{Homogeneity} = \sum_i \sum_j \frac{1}{1 + (i-j)^2} P(i,j) \tag{5}$$

$$\text{Variance} = \sum_i \sum_j (i - \mu)^2 P(i,j) \tag{6}$$

$$\text{Mean} = \sum_{i=2}^{2N_g} iP_{x+y}(i) \tag{7}$$

$$\text{Entropy} = -\sum_i P \log(P(i,j)) \tag{8}$$

$$\text{Dissimilarity} = \sum_{i,j=0}^{N_g-1} P_{i,j}(-\ln P_{i,j}) \tag{9}$$

where $p(i, j)$ is the $(i, j)$-th entry in a normalized gray-tone spatial dependence matrix $P(i, j)/R$; $R$ is the total sum of $P$; $p_x(i) = \sum_{j=1}^{Ng} P(i, j)$ is the $i$-th entry in the marginal probability matrix obtained by summing the rows of $p(i, j)$; and $\mu_x$, $\mu_y$, $\sigma_x$ and $\sigma_y$ are the means and standard deviations of $p_x$ and $p_y$.

In this study, eight textural features at angle $0°$ and distance 1, different window sizes ranging from $3 \times 3$ to $21 \times 21$ and a quantization level of 64 were used to evaluate its performance for classification.

## 3.2. Principal Component Analysis

A set of eight texture features were calculated for each scattering element, and a total of 32 texture measures were obtained from the ALOS-2 and Sentinel-1 images. Performing land cover classification from these high dimensional datasets could be inefficient and time consuming. Thus, to reduce the dimensionality, the principal component analysis (PCA) was performed independently on each set of eight texture features. Since PCA computes the correlation between input bands and sorts them based on the amount of data variance [35], the first components contain the greatest variance of the input variables [36]. In this research, three of the first components were selected, which contained almost 99 percent of the variation of the input elements and were used for the next stage. The reason on behalf the use of the first three principal components is explained further in Section 4.2. Therefore, two sets of SAR data were used for the supervised classification: (1) the backscatter values of the dual-polarization data, which contains two layers of HH + HV and VV + VH for the ALOS-2 and Sentinel-1; and (2) the layer stacking of the backscatter values and the first three principal components of the texture measures (PCT). The second dataset includes eight layers (HH, HV, $PCT_{1,2,3}^{HH}$, $PCT_{1,2,3}^{HV}$) for ALOS-2 and eight layers (VV, VH, $PCT_{1,2,3}^{VV}$, $PCT_{1,2,3}^{VH}$) for Sentinel-1.

*3.3. Supervised Classification*

Supervised classification is a training based methodology that classifies similar image pixel values into training samples for a determined number of classes. Thus, training samples must be selected based on a homogenous group of image pixels to provide the best separability. Therefore, monitoring the study area and assessing the different land covers is necessary before selecting the training samples. Additionally, applying the appropriate algorithm to identify the homogeneity of the training data to group the pixel values of a dataset is important. Accordingly, the training data selection process and the two different supervised classification algorithms used will be explained further in this section.

After the inspection of district 22 in Tehran, five land cover classes were defined: bare land, vegetation, built-up 1, built-up 2, and built-up 3. The three different residential types are shown in Figure 3. Built-up 1 is composed of dense residential areas with mostly two-story old buildings and narrow streets and roads. The buildings are approximately 8 × 13 m in size. Built-up 2 is composed of medium density residential areas, including parallel blocks of buildings. The buildings are mostly 12 × 18 m in size. These buildings consist of approximately 4–6 floors. Built-up 3 includes individual building with 4–15 floors. Built-up 3 has wider streets than the other areas and plenty of vegetation surrounding the buildings. The buildings are mostly two different shapes, square with a size of 25 × 25 m or rectangular with a size of approximately 25 × 44 m. For each of these five land cover classes, three training samples were selected. The training samples mainly consist of square shapes of 200 m in length, which are located in different parts of Tehran, as shown in Figure 4.

The classification process was performed using two algorithms: support vector machine (SVM) and maximum likelihood (ML). Both algorithms are derived from statistical theories. These two methods are commonly used in land cover classification studies [37–41]; therefore, we intend to evaluate their performance for SAR data imagery. Supervised classifiers were applied to three sets of data: (1) Landsat 8 multi-spectral image; (2) Backscatter values of dual-polarization ALOS-2 and Sentinel-1 data; and (3) the layer stacking of backscatter values and the first three PCTs from ALOS-2 and Sentinel-1 data.

(a) Built-up 1     (b) Built-up 2     (c) Built-up 3

**Figure 3.** Three different residential types (a–c) extracted from Google Earth images of Tehran city.

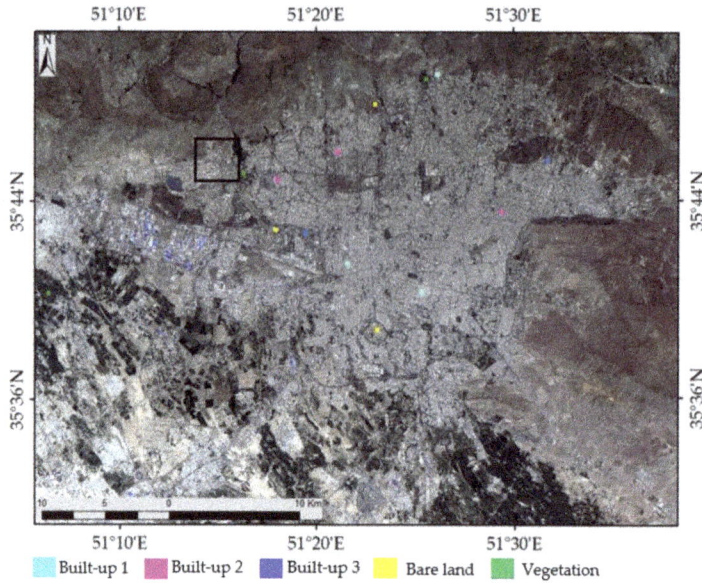

**Figure 4.** The three training samples selected for each land cover class from the Landsat 8 image used for supervised classification. The black frame shows the area used for assessing the accuracy.

The maximum likelihood method works based on the assumption that each class is normally distributed. Thus, for each pixel, the probability that it belongs to a specific class is calculated. Then, the pixel is assigned to the class that yields the highest probability as follows [36,42]:

$$g_i(x) = 1np(\omega_i) - 1/21n\left|\sum_i\right| - 1/2(x - m_i)^T \sum_i^{-1}(x - m_i) \tag{10}$$

where $i$ is the number of classes, $x$ represents the $n$-dimensional data (where $n$ is the number of bands), $p(\omega_i)$ denotes the probability that class $\omega_i$ occurs in the image and is assumed to be the same for all classes, $|\Sigma_i|$ is the determinate of the covariance matrix of the data in class $\omega_i$, $\Sigma_i^{-1}$ is its inverse matrix, and $m_i$ represents the mean vector.

Support vector machine is a trusted algorithm often used in remote sensing [43,44]. This algorithm was developed by Vanpik [45], and its use has increased for land cover classification in recent years. The SVM separates the pixels of an image using optimal hyperplanes that maximize the margin between the classes [46]. The data points closest to a hyperplane are called support vectors. A nonlinear classification can be performed using kernel functions to the support vectors. In this study, the pairwise strategy is used for multi-class classification. We selected the radial basis function, which is a common kernel type one for the classification and is expressed as follows:

$$K(x_i, x_j) = \exp\left(-g\|x_i - x_j\|^2\right) \quad, g > 0 \tag{11}$$

where $x_i$ and $x_j$ represent the training data and class labels, and $g$ is the gamma term in the kernel function [47–49]. Radial basis functions with gamma value of 1.12 and 1.14 were used for SAR and optical image, respectively. The gamma value was calculated by inversing of the number of bands in the input image.

Moreover, the memory usage and the speed of training were not considered for either supervised algorithms in this study because the scope of this research is to evaluate the improvement of the accuracy after introducing the texture feature.

*3.4. Accuracy Assessment*

The accuracy of the results from the ML and SVM methods was measured by calculating a confusion matrix. The confusion matrix compares the classified land cover with truth data and is a standard method used to evaluate the accuracy of classification in remote sensing [50]. The overall accuracy, producer and user accuracies, and kappa coefficient are calculated from this matrix.

In this research, the confusion matrix was prepared using truth data over the five land cover classes. Based on the comparison of the confusion matrix of the classification results, we evaluated the effect of the supervised classification algorithms and texture measures.

## 4. Results and Discussion

This study aims to classify the land covers in Tehran city with high accuracy using appropriate datasets and methodologies. A multi-spectral Landsat 8 image and SAR data (ALOS-2 and Sentinel-1) were used to evaluate the performance of the supervised classification algorithms, maximum likelihood and support vector machine. The comparison and evaluation of the results from the optical and SAR sensors are presented in the following.

*4.1. Multi-Spectral Optical Image*

Regarding the aim of this study, we begin by observing the performance of the multi-spectral optical image. First, the spectral signatures of the different land cover samples from the Landsat 8 image (Figure 4) for the built-up 1, built-up 2, built-up 3, bare land, and vegetation classes are shown in Figure 5. The spectral reflectance characteristics of bare lands and built-up features are remarkably similar, while the vegetation signature shows a different pattern. The geography of the study area makes it difficult to differentiate the urban area from the mountainous and desert areas that surround Tehran. ML and SVM classifications were applied to the Landsat 8 image using a total of fifteen training samples from five different land cover types. The results are shown in Figure 6. The ML classification result (Figure 6a) indicates that almost all bare land and mountain areas around Tehran were selected as built-up area (built-up 1). Additionally, the SVM result (Figure 6b) shows that some bare land areas in the southeast and west of Tehran were classified incorrectly as built-up areas (built-up 1 and 2).

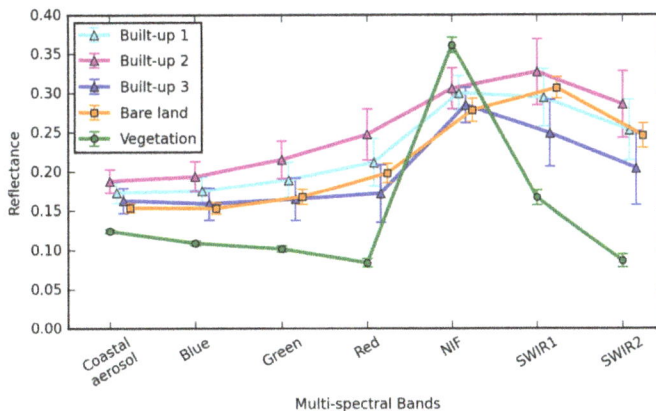

**Figure 5.** Spectral signatures of the five land covers obtained from the Landsat 8 image.

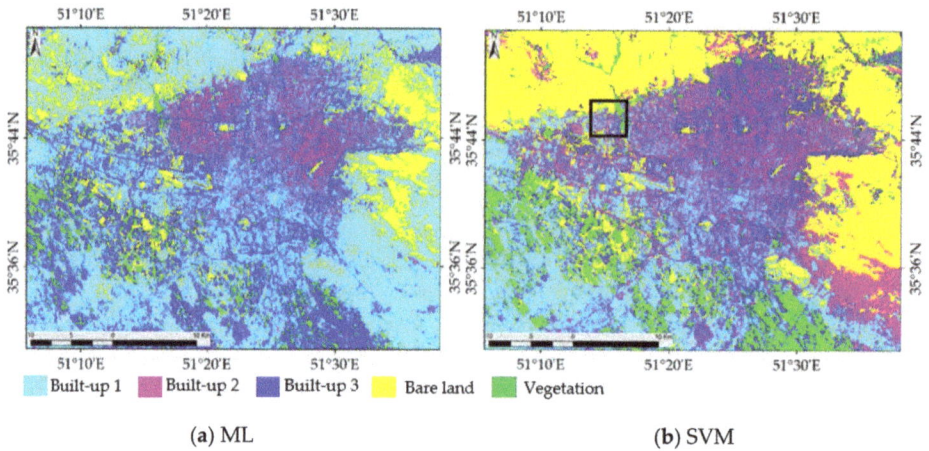

51°10'E    51°20'E    51°30'E

Built-up 1   Built-up 2   Built-up 3   Bare land   Vegetation

(a) ML          (b) SVM

**Figure 6.** The ML (**a**) and SVM (**b**) classification results of Landsat 8 image, where the black frame shows the location of the validation area.

To validate the classification result, we prepared the confusion matrix using the truth data. Figure 7a represents a closer look of the validation area, which is shown by a black frame in Figures 4 and 6b as well. Table 2 illustrates the confusion matrix results from Landsat 8 using SVM and ML classification. The SVM classification gave an overall accuracy and kappa coefficient of 41.1% and 0.26, respectively, while ML gave the values of 35.3% and 0.22. Moreover, the producer and user accuracy are shown in table below. The producer accuracy represents the correctly classified pixels out of the truth pixels of related land cover classes. The user accuracy, illustrates the correctly classified pixels out of the total classified pixels. The SVM algorithm provided higher accuracy than ML. A closer look at the SVM result is shown in Figure 7b. It is observed that some classes were incorrectly classified, such as built-up 2, which was mostly classified as built-up 1. Therefore, the Landsat 8 image does not seem appropriate for classifying the land cover of the study area, Tehran city.

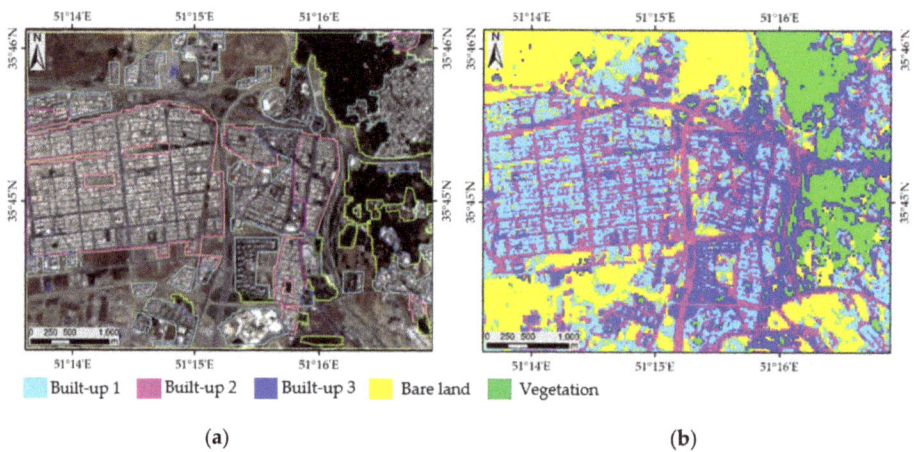

Built-up 1   Built-up 2   Built-up 3   Bare land   Vegetation

(a)          (b)

**Figure 7.** Validation area containing the truth data for the five classes (**a**); and the result of the SVM classification from Landsat 8 image (**b**).

**Table 2.** Confusion matrix for ML and SVM classification using the Landsat 8 image.

| | Land Cover Class | Bare Land | Vegetation | Built-up 1 | Built-up 2 | Built-up 3 | Overall Accuracy (%) | Kappa Coefficient |
|---|---|---|---|---|---|---|---|---|
| ML | Producer Accuracy (%) | 30.7 | 58.7 | 31.5 | 35.0 | 33.0 | 35.3 | 0.22 |
| | User Accuracy (%) | 91.2 | 81.4 | 4.0 | 47.8 | 22.0 | | |
| SVM | Producer Accuracy (%) | 42.4 | 63.8 | 31.2 | 39.8 | 27.2 | 41.1 | 0.26 |
| | User Accuracy (%) | 86.4 | 76.4 | 5.1 | 39.3 | 25.5 | | |

## 4.2. SAR Data

Considering the limitation of the Landsat 8 image in land cover classification and detection of various built-up classes and bare land in the study area, ALOS-2 and Sentinel-1 images with the capability of obtaining ground surface information based on the backscatter coefficient were selected to overcome this issue. Figure 8 represents the color composites of the dual-polarization intensity images from the ALOS-2 and Sentinel-1 data. The cyan and red colors indicate different orientation and geometrical positions of the residential areas in Tehran.

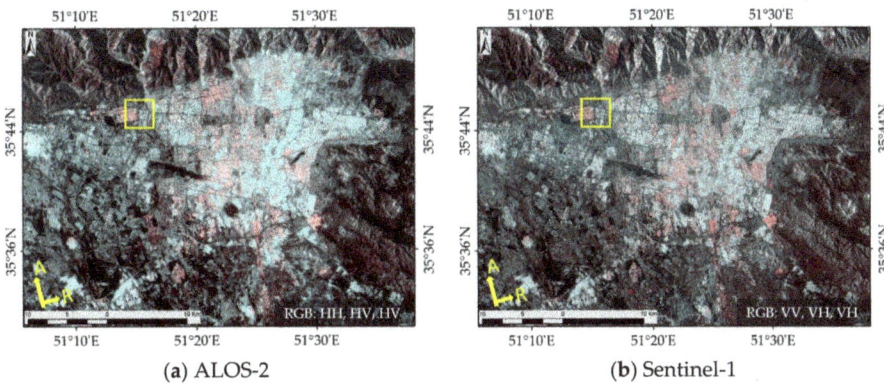

(a) ALOS-2          (b) Sentinel-1

**Figure 8.** Color composite of dual-polarization images from the ALOS-2 image (**a**) and the Sentinel-1 image (**b**), the yellow frame shows the location of region chosen for evaluating the components of principal component analysis (PCA) and the location of validation area.

To increase the accuracy in the supervised classification from SAR data, texture analysis (GLCM) was applied to the dual-polarization images. Figure 9 depicts the texture features obtained from ALOS-2 HH polarization. Texture analysis was applied to each backscatter value of HH, HV, VV, and VH. Thus, for each polarized image, eight texture measures were obtained: mean, contrast, ASM, correlation, homogeneity, variance, entropy, and dissimilarity. Then, the PCA was applied to reduce the number of textures measured.

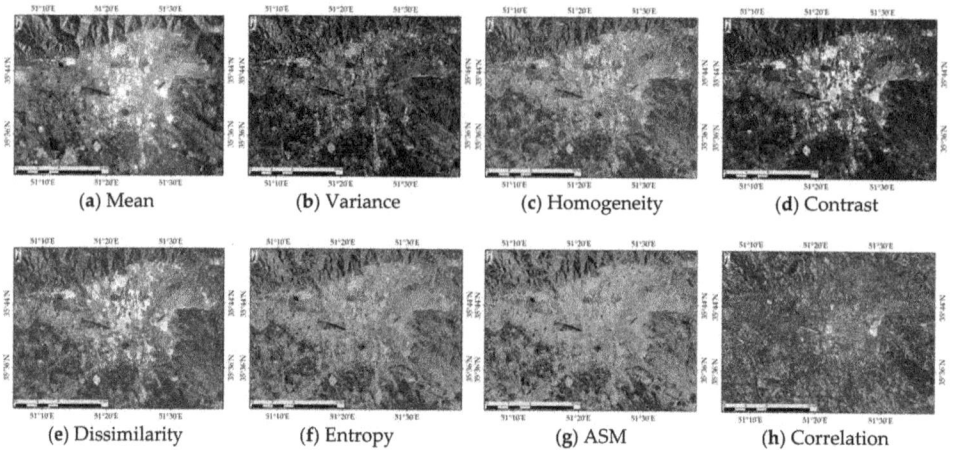

**Figure 9.** Eight texture features (**a–h**) obtained from the ALOS-2 HH polarization with a window size of 13 × 13, angle of 0°, and displacement of 1 pixel.

In order to select the appropriate number of principal components for texture measures (PCT) to perform the classification, we examined the region shown as a yellow frame in Figure 8 using the HH and HV polarizations of ALOS-2. The SVM classification was carried out for two datasets. In the first dataset, the two backscatter values (HH and HV) and all their PCTs (18 layers in total) were used. In the second dataset, the two backscatter values and their first three PCTs (eight layers in total) were used. In both datasets the PCT was calculated from texture features obtained using a window size of 13 × 13. The first three components contain almost 99 percent of the variation of the input elements. Table 3 shows the overall accuracy and the kappa coefficient obtained from the classification of the two datasets. A comparison shows a difference of only 1.26% for the overall accuracy and a difference of 0.02 for the kappa coefficient. Although the dataset with all the PCTs gained higher accuracy, the required time was significantly larger than that of the second dataset. Therefore, in this study, only the first three principal components were used.

**Table 3.** Overall accuracy and kappa coefficient calculated from the SVM classification using the first three PCTs and all PCTs.

| Datasets | Overall Accuracy (%) | Kappa Coefficient |
|---|---|---|
| Backscatter + three PCTs (%) | 61.79 | 0.50 |
| Backscatter + all PCTs (%) | 63.05 | 0.52 |

Figure 10 illustrates the first three PCTs of the HH polarization. The classifications using ML and SVM were performed using the same fifteen training samples shown in Figure 4. As mentioned before, the performance of both ML and SVM are compared to the truth data to observe which method produces the highest accuracy.

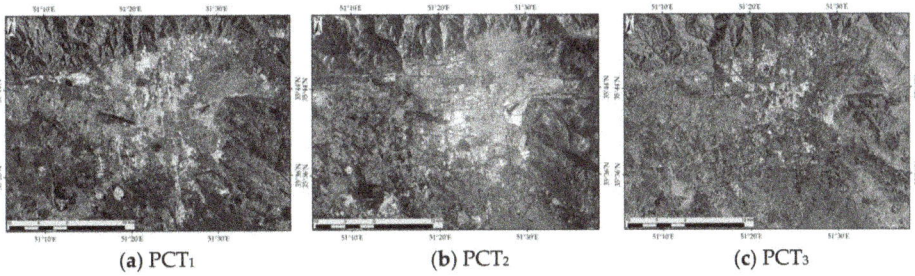

**Figure 10.** The first three principal components (**a–c**) prepared from the eight texture features of the ALOS-2 HH polarization.

The window size is an important parameter for the texture measures. In this study, window sizes ranging from 3 × 3 to 21 × 21 were all evaluated to estimate the most suitable window size. Figure 11 shows the kappa coefficients for the classification using only the backscatter values of ALOS-2 (HH, HV) and Sentinel-1 (VV, VH), and those by the layer stacking of the backscatter values and their PCTs results for different window sizes. The numbers of pixels in each land cover class in the truth data are not equal, thus the kappa coefficient is plotted in Figure 11. The solid lines represent the results from SVM classification and the dashed lines those from ML classification, both for ALOS-2 and Sentinel-1. The graph shows an increase of the kappa coefficient by introducing the texture measures for the classification. The performance of SVM is better than that of ML classification. For the ALOS-2 satellite, the difference between the both methods increases significantly when the texture window size increases. The highest kappa coefficient was observed in the classification for ALOS-2 with window size 13 by SVM as shown by green arrow in the graph. The highest kappa coefficient of Sentinel-1 classification results was obtained from SVM in window size 11 × 11. Moreover, ALOS-2 provided better performance than Sentinel-1.

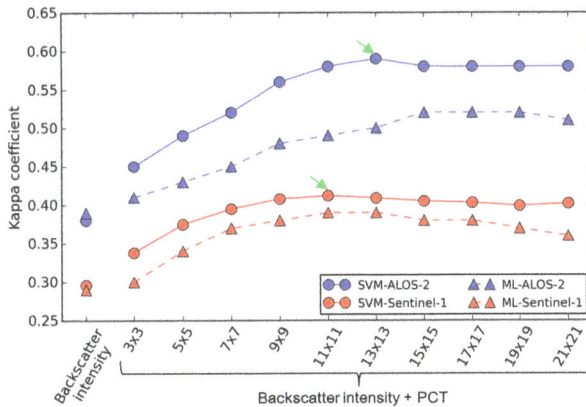

**Figure 11.** Kappa coefficient calculated from SVM and ML classification methods using only the backscatter of ALOS-2 and Sentinel-1 and those by the layer stacking of the backscatter values and their PCTs for window sizes ranging from 3 × 3 to 21 × 21. Green arrows show the highest Kappa coefficient.

Figure 12 illustrates the SVM classification of Tehran using the layer stacking of backscatter values and their PCTs for the both ALOS-2 and Sentinel-1 satellites. The PCT was calculated using a window sizes of 13 × 13 and 11 × 11 for the ALOS-2 and Sentinel-1, respectively.

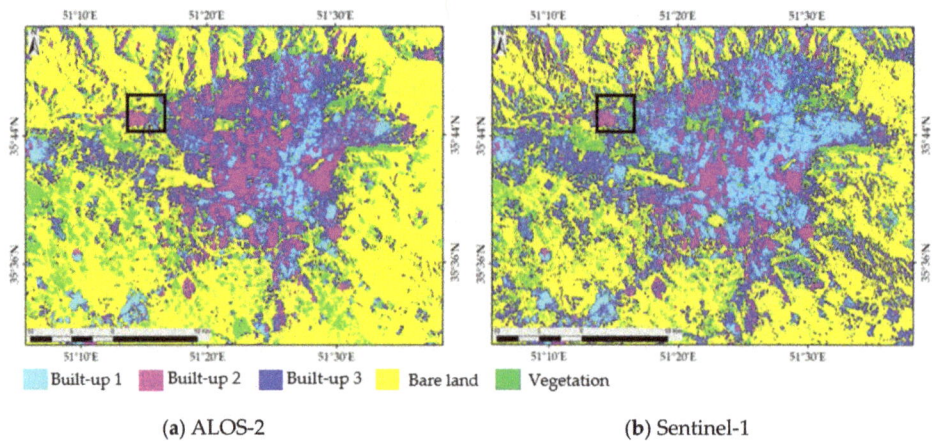

Built-up 1   Built-up 2   Built-up 3   Bare land   Vegetation

(a) ALOS-2          (b) Sentinel-1

**Figure 12.** SVM classification results of backscatter values and their PCTs from the ALOS-2 image using a window size of 13 × 13 (**a**) and the Sentinel-1 image using a window size of 11 × 11 (**b**). The black frame represents the location of Figure 13.

Herein, further evaluation is performed only for the datasets that produced the highest kappa coefficient. Thus, for the sake of brevity, the classification result from the ALOS-2 backscatter values and their first three PCTs using a window size of 13 × 13 is referred as ALOS-2-PCT. Similarly, the classification result from the Sentinel-1 backscatter values and their first three PCTs using a window size of 11 × 11 is referred to as Sentinel-1-PCT. Tables 4–6 illustrate the confusion matrix for the Landsat-8 image, ALOS-2 (backscatter values only), and ALOS-2-PCT, respectively. Tables 7 and 8 represent the confusion matrix for the Sentinel-1, and the Sentinel-1 PCT, respectively. The tables show a remarkable improvement in both overall accuracy and kappa coefficient for the datasets that include texture measures over the datasets of only backscattering values and Landsat 8. The Landsat 8 produced an overall accuracy and kappa coefficient of 41.1% and 0.25, respectively. Tables 5 and 6 depict that the overall accuracy and kappa coefficient of the ALOS-2-PCT is greater than the ALOS-2. The overall accuracy increased from 53.0% to 69.7% and the kappa coefficient from 0.38 to 0.58 when the texture measures of window size 13 × 13 were included. Moreover, the Sentinel-1 PCT shows higher overall accuracy and kappa coefficient than the Sentinel-1. Tables 7 and 8 show an increase from 45.7% to 54.2% for the overall accuracy and an increment from 0.29 to 0.41 for the kappa coefficient when the textures of window size 11 × 11 were included.

The confusion matrix includes the producer and user accuracies as well. The diagonal elements in these tables depict the correctly classified pixels in each land cover class. The comparison shows that the highest producer accuracy was obtained when the texture features were included (Tables 6 and 8). It can be observed from Tables 4 and 5 that, although the producer accuracy increased for the bare land, built-up 1 and built-up 2 classes when the SAR backscatter values in classification is used instead of the Landsat 8, the producer accuracy decreased for the built-up 3 and vegetation classes from 27.1% and 63.8% to 15.6% and 59.6%. However, when texture features are included, an improvement is observed for all the land cover classes. Thus, in our study area, classification of land cover using SAR backscatter and texture feature is superior in terms of accuracy than classification from the Landsat 8 and SAR backscatter only. Furthermore, the user accuracy increased in all classes except vegetation for the ALOS-2-PCT. In case of the Sentinel-1, the user accuracy for the dataset including the texture measures improved in all the classes comparing with the dataset of backscattering values only and there was no improvement for the bare land and vegetation comparing with the Landsat 8 dataset.

**Table 4.** Confusion matrix for SVM classification using the Landsat 8 image.

| | | Land Cover Classes of Truth Data | | | | | | User Accuracy (%) |
|---|---|---|---|---|---|---|---|---|
| | | Bare land | Built-up 1 | Built-up 2 | Built-up 3 | Vegetation | Total | |
| Land Cover Classification from Satellite | Bare land | 15,044 | 159 | 728 | 1094 | 380 | 17,405 | 86.43 |
| | Built-up 1 | 6057 | 1130 | 9677 | 4807 | 367 | 22,038 | 5.13 |
| | Built-up 2 | 7598 | 952 | 8335 | 3866 | 473 | 21,224 | 39.27 |
| | Built-up 3 | 5556 | 1184 | 2132 | 3771 | 2148 | 14,791 | 25.50 |
| | Vegetation | 1221 | 199 | 68 | 347 | 5943 | 7778 | 76.41 |
| | Total | 35,476 | 3624 | 20,940 | 13,885 | 9311 | 83,236 | |
| Producer Accuracy (%) | | 42.41 | 31.18 | 39.80 | 27.16 | 63.83 | | 41.12 |
| Kappa Coefficient 0.2594 | | | | | | | | |

**Table 5.** Confusion matrix for SVM classification using the ALOS-2 (backscattering values only).

| | | Land Cover Classes of Truth Data | | | | | | User Accuracy (%) |
|---|---|---|---|---|---|---|---|---|
| | | Bare land | Built-up 1 | Built-up 2 | Built-up 3 | Vegetation | Total | |
| Land Cover Classification from Satellite | Bare land | 105,005 | 239 | 8719 | 6705 | 3098 | 123,766 | 84.84 |
| | Built-up 1 | 3804 | 10,126 | 10,754 | 12,052 | 5648 | 42,384 | 23.89 |
| | Built-up 2 | 7603 | 1707 | 65,115 | 11,215 | 2772 | 88,412 | 73.65 |
| | Built-up 3 | 20,145 | 730 | 11,703 | 10,046 | 6604 | 49,228 | 20.41 |
| | Vegetation | 40,800 | 4290 | 9026 | 24,242 | 26,833 | 105,191 | 25.51 |
| | Total | 177,357 | 17,092 | 105,317 | 64,260 | 44,955 | 408,981 | |
| Producer Accuracy (%) | | 59.21 | 59.24 | 61.83 | 15.63 | 59.69 | | 53.08 |
| Kappa Coefficient 0.3840 | | | | | | | | |

**Table 6.** Confusion matrix for SVM classification using the ALOS-2-PCT.

| | | Land Cover Classes of Truth Data | | | | | | User Accuracy (%) |
|---|---|---|---|---|---|---|---|---|
| | | Bare land | Built-up 1 | Built-up 2 | Built-up 3 | Vegetation | Total | |
| Land Cover Classification from Satellite | Bare land | 133,759 | 419 | 5886 | 10,340 | 3993 | 154,397 | 86.63 |
| | Built-up 1 | 409 | 11,285 | 11,886 | 4093 | 631 | 29,304 | 38.51 |
| | Built-up 2 | 1805 | 319 | 66,121 | 4656 | 570 | 73,471 | 90.00 |
| | Built-up 3 | 26,246 | 3265 | 20,458 | 42,277 | 7996 | 100,242 | 42.17 |
| | Vegetation | 14,138 | 1804 | 966 | 2894 | 31,765 | 51,567 | 61.60 |
| | Total | 177,357 | 17,092 | 105,317 | 64,260 | 44,955 | 408,981 | |
| Producer Accuracy (%) | | 75.42 | 66.03 | 62.78 | 65.79 | 70.66 | | 69.73 |
| Kappa Coefficient 0.5881 | | | | | | | | |

**Table 7.** Confusion matrix for SVM classification using the Sentinel-1 (backscattering values only).

| | | Land Cover Classes of Truth Data | | | | | | User Accuracy (%) |
|---|---|---|---|---|---|---|---|---|
| | | Bare land | Built-up 1 | Built-up 2 | Built-up 3 | Vegetation | Total | |
| Land Cover Classification from Satellite | Bare land | 88,767 | 735 | 11,727 | 11,954 | 7738 | 120,921 | 73.41 |
| | Built-up 1 | 7713 | 8518 | 13,895 | 14,617 | 4472 | 49,215 | 17.31 |
| | Built-up 2 | 12,426 | 3142 | 62,049 | 13,321 | 2900 | 93,838 | 66.12 |
| | Built-up 3 | 1954 | 1155 | 915 | 2120 | 1910 | 8054 | 26.32 |
| | Vegetation | 68,137 | 4235 | 18,317 | 25,604 | 29,546 | 145,839 | 20.26 |
| | Total | 178,997 | 17,785 | 106,903 | 67,616 | 46,566 | 417,867 | |
| Producer Accuracy (%) | | 49.59 | 47.89 | 58.04 | 3.14 | 63.45 | | 45.70 |
| Kappa Coefficient 0.2963 | | | | | | | | |

**Table 8.** Confusion matrix for SVM classification using the Sentinel-1-PCT.

| | | Land Cover Classes of Truth Data | | | | | | User Accuracy (%) |
|---|---|---|---|---|---|---|---|---|
| | | Bare land | Built-up 1 | Built-up 2 | Built-up 3 | Vegetation | Total | |
| Land Cover Classification from Satellite | Bare land | 85,695 | 97 | 5082 | 4392 | 4101 | 99,367 | 86.24 |
| | Built-up 1 | 2250 | 11,060 | 14,621 | 11,330 | 1130 | 40,391 | 27.38 |
| | Built-up 2 | 3772 | 431 | 61,405 | 6131 | 194 | 71,933 | 85.36 |
| | Built-up 3 | 61,738 | 4387 | 23,477 | 38,180 | 10,779 | 138,561 | 27.55 |
| | Vegetation | 25,542 | 1810 | 2318 | 7583 | 30,362 | 67,615 | 44.90 |
| | Total | 178,997 | 17,785 | 106,903 | 67,616 | 46,566 | 417,867 | |
| Producer Accuracy (%) | | 47.88 | 62.19 | 57.44 | 56.47 | 65.20 | | 54.25 |
| Kappa Coefficient 0.4122 | | | | | | | | |

Figure 13 shows a closer look of SVM classification for the ALOS-2, Sentinel-1, ALOS-2-PCT, and Sentinel-1-PCT. The location of the area is shown in Figure 12. The improvement in the accuracy mentioned before can be appreciated visually in this figure. The most remarkable observation is that when using texture measures of the SAR data for classification, the producer accuracy improved significantly, therefore, the amount of noise and misclassified pixels for all classes decreased and the classes become more uniform comparing to that using only SAR backscatter values or optical data.

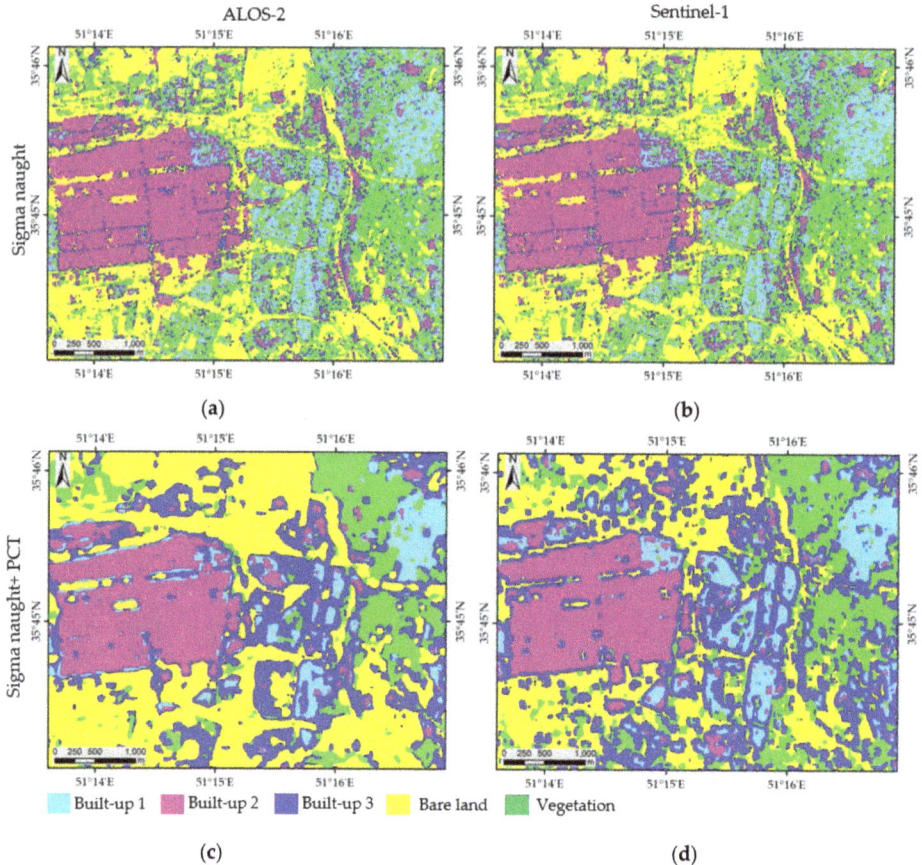

**Figure 13.** A close-up of the SVM classification results from the ALOS-2 (**a**); Sentinel-1 (**b**); ALOS-2 PCT (**c**); and Sentinel-1 PCT (**d**).

## 5. Conclusions

In this study, the GLCM texture measures were applied to improve the supervised classification of SAR intensity images for urban areas. For this purpose, Tehran was selected as the study area because of its rapid expansion, which has resulted in land cover changes and the appearance of new urban regions. Due to the similarity of the spectral signatures of soil and roof material in built-up regions, classification from multi-spectral optical images seems very difficult. Alternatively, SAR images may be a better option because the backscattering depends on the geometrical features of the objects within the recorded area. Dual-polarized data from L-band ALOS-2 (HH, HV) and C-band Sentinel-1 (VV, VH) were employed. In addition, the texture properties were calculated by applying a gray-level co-occurrence matrix (GLCM). Thus, eight texture features were obtained for each intensity element. Furthermore, a principal component analysis was applied to each set of texture measures, and the first three components were selected based on the greatest covariance. Then, maximum likelihood and support vector machine algorithms were used for the three datasets: optical images and the SAR intensity data without and with texture measures. The results of the supervised classification were compared with the truth data obtained by visual inspection of the Landsat 8 image.

The supervised classification results with texture measures were found to be superior to the results without texture in two main aspects: the highest accuracy and least noise. The support vector machine for both the optical and SAR sensors produced a higher accuracy than the maximum likelihood. Moreover, the classification of ALOS-2 with the SVM methodology using a window size of $13 \times 13$ obtained the highest overall accuracy. Besides, Sentinel-1 gained the best accuracy in window size $11 \times 11$ with SVM classification. Although the intermediate window sizes of $13 \times 13$ and $11 \times 11$ worked well in this study area, the best window size could change based on the different study areas and truth data using for assessing the accuracy.

**Acknowledgments:** The ALOS-2 PALSAR-2 data used in this study are owned by the Japan Aerospace Exploration Agency (JAXA) and were provided through the JAXA's ALOS-2 research program (RA4, PI No. 1503). The Pleiades images are owned by Airbus Defense and Space and licensed to Chiba University.

**Author Contributions:** This research was conducted by Homa Zakeri for her Doctoral studies under the supervision of Fumio Yamazaki and guidance of Wen Liu. All authors read and reviewed the article.

**Conflicts of Interest:** The authors declare no conflict of interest.

## References

1. Bhatta, B. *Advanced in Geographic Information Science: Analysis of Urban Growth and Sprawl from Remote Sensing Data*; Springer: Heidelberg, Germany, 2010; pp. 1–167.
2. Brando, V.E.; Dekker, A.G. Satellite hyperspectral remote sensing for estimating estuarine and coastal water quality. *IEEE Trans. Geosci. Remote Sens.* **2003**, *41*, 1378–1387. [CrossRef]
3. Molch, K. *Radar Earth Observation Imagery for Urban Area Characterisation*; JRC Scientific and Technical Reports; European Commission: Luxembourg, 2009; pp. 1–19.
4. Turner, B.L.; Lambin, E.F.; Reenberg, A. The emergence of land change science for global environmental change and sustainability. *Proc. Natl. Acad. Sci. USA* **2007**, *104*, 20666–20671. [CrossRef] [PubMed]
5. Rathje, E.; Adams, B.J. The Role of Remote Sensing in Earthquake Science and Engineering: Opportunities and Challenges. *Earthq. Spectra* **2008**, *24*, 471–492. [CrossRef]
6. Yamazaki, F.; Inoue, H.; Liu, W. Characteristics of SAR backscattering intensity and its application to earthquake damage detection. In Proceedings of the 6th Conference on Computational Stochastic Mechanics, Rodhes, Greece, 13–16 June 2010; pp. 602–606.
7. Gong, P.; Wang, J.; Yu, L.; Zhao, Y.; Zhao, Y.; Liang, L.; Niu, Z.; Huang, X.; Fu, H.; Liu, S.; et al. Finer resolution observation and monitoring of global land cover: First mapping results with Landsat TM and ETM+ data. *Int. J. Remote Sens.* **2012**, *34*, 2607–2654. [CrossRef]
8. Liu, W.; Yamazaki, F.; Goken, H.; Koshimura, S. Extraction of Tsunami-Flooded Areas and Damaged Buildings in the 2011 Tohoku-Oki Earthquake from TerraSAR-X Intensity Images. *Earthq. Spectra* **2013**, *29* (Suppl. 1), S183–S200. [CrossRef]
9. Yamazaki, F.; Liu, W. Urban change monitoring: Multi-temporal SAR images. In *Encyclopedia of Earthquake Engineering*; Beer, M., Kougioumtzoglou, I., Au, S., Eds.; Springer: Berlin/Heidelberg, Germany, 2015; pp. 3847–3859.
10. Guan, D.; Li, H.; Inohae, T.; Su, W.; Nagaie, T.; Hokao, K. Modeling urban land use change by the integration of cellular automaton and Markov model. *Ecol. Model.* **2011**, *222*, 3761–3772. [CrossRef]
11. Alqurashi, A.F.; Kumar, L.; Sinha, P. Urban Land Cover Change Modelling Using Time-Series Satellite Images: A Case Study of Urban Growth in Five Cities of Saudi Arabia. *Remote Sens.* **2016**, *8*, 838. [CrossRef]
12. Zhang, J.; Li, P.; Wang, J. Urban built-Up area extraction from landsat TM/ETM+ images using spectral information and multivariate texture. *Remote Sens.* **2014**, *6*, 7339–7359. [CrossRef]
13. Zhang, Y. Optimisation of building detection in satellite images by combining multispectral classification and texture filtering. *Photogramm. Remote Sens.* **1999**, *54*, 50–60. [CrossRef]
14. Colaninno, N.; Roca, J.; Burns, M.; Alhaddad, B.; Patterns, U.; Management, A. Defining densities for urban residential texture, through land use classification, from landsat TM imagery: Case Study of Spanish

Mediterranean Coast. In Proceedings of the XXII Congress of the ISPRS Congress, Melbourne, Australia, 25 August–1 September 2012; pp. 179–184.

15. Henderson, F.M.; Lewis, A.J. Introduction. In *Principles and Applications of Imaging Radar*, 3rd ed.; Manual of Remote Sensing; John Wiley & Sons Inc.: New York, NY, USA, 1998; pp. 1–6.

16. Lillesand, T.M.; Kiefer, R.W.; Chipman, J.W. *Remote Sensing and Image Interpretation*, 5th ed.; John Wiley & Sons Inc.: New York, NY, USA, 2004; pp. 638–713.

17. Rogan, J.; Chen, D. M. Remote sensing technology for mapping and monitoring Land-cover and land-use change. *Prog. Plan.* **2004**, *61*, 301–325. [CrossRef]

18. Lee, J.S.; Grünes, M.R.; Ainsworth, T.L.; Du, L.J.; Schuler, D.L.; Cloude, S.R. Unsupervised classification using polarimetric decomposition and the complex wishart classifier. *IEEE Trans. Geosci. Remote Sens.* **1999**, *37 Pt 1*, 2249–2258.

19. Hoekman, D.H. A new polarimetric classification approach evaluated for agricultural crops. *Eur. Space Agency* **2003**, *41*, 71–79. [CrossRef]

20. Dell'Acqua, F.; Gamba, P. Texture-based characterization of urban environmental on satellite SAR images. *IEEE Trans. Geosci. Remote Sens.* **2003**, *41*, 153–159. [CrossRef]

21. Matsuoka, M.; Yamazaki, F. Use of satellite SAR intensity imagery for detecting building areas damaged due to earthquakes. *Earthq. Spectra* **2004**, *20*, 975–994. [CrossRef]

22. Haralick, R.; Shanmugan, K.; Dinstein, I. Textural features for image classification. *IEEE Trans. SMC* **1973**, *3*, 610–621. [CrossRef]

23. Augusteijn, M.F.; Clemens, L.E.; Shaw, K.A. Performance evaluation of texture measures for ground cover identification in satellite images by means of a neural network classifier. *IEEE Trans. Geosci. Remote Sens.* **1995**, *33*, 616–626. [CrossRef]

24. Shaban, M.A.; Dikshit, O. Improvement of classification in urban areas by the use of textural features: The case study of Lucknow City, Uttar Pradesh. *Int. J. Remote Sens.* **2010**, *22*, 565–593. [CrossRef]

25. Lu, D.; Batistella, M.; Moran, E. Integration of Landsat TM and SPOT HRG Images for Vegetation Change Detection in the Brazilian Amazon. *Photogramm. Eng. Remote Sens.* **2008**, *74*, 421–430. [CrossRef]

26. Clausi, D.A.; Yu, B. Comparing co-occurrence probabilities and Markov random fields for texture analysis of SAR Sea ice imagery. *IEEE Trans. Geosci. Remote Sens.* **2004**, *42*, 215–228. [CrossRef]

27. Kandaswamy, U.; Adjeroh, D.A.; Lee, M.C. Efficient Texture Analysis of SAR Imagery. *IEEE Trans. Geosci. Remote Sens.* **2005**, *43*, 2075–2083. [CrossRef]

28. Atlas of Tehran Metropoli. Available online: http://atlas.tehran.ir/Default.aspx?tabid=227 (accessed on 11 February 2017).

29. Madanipour, A. Urban planning and development in Tehran. *Cities* **2006**, *23*, 433–438. [CrossRef]

30. District 22 of Tehran Municipality. Available online: http://region22.tehran.ir/ (accessed on 11 February 2017).

31. Advanced Land Observing Satellite DAICHI-2. DAICHI. Available online: http://www.eorc.jaxa.jp/ALOS/en/top/about_top.htm (accessed on 11 February 2017).

32. Sentinel Online. Available online: https://sentinel.esa.int/web/sentinel/missions/sentinel-1/instrument-payload (accessed on 11 February 2017).

33. Lee, J.S.; Grunes, M.R.; de Grandi, G. Polarimetric SAR Speckle Filtering and Its Implication for Classification. *IEEE Trans. Geosci. Remote Sens.* **1999**, *37*, 2363–2373.

34. Ulaby, F.T.; Moore, R.K.; Fung, A.K. *Microwave Remote Sensing: Active and Passive, Radar Remote Sensing and Surface Scattering and Emission Theory*, 82nd ed.; Addison Wesley: Boston, MA, USA, 1982; Volume 2, pp. 1–624.

35. Anyamba, A.; Eastman, J.R. Interannual variability of NDVI over Africa and its relation to El Nino/Southern Oscillation. *Int. J. Remote Sens.* **1996**, *17*, 2533–2548. [CrossRef]

36. Richards, J.A. *Remote Sensing Digital Image Analysis*, 5th ed.; Springer: Berlin/Heidelberg, Germany, 2013; pp. 192–195.

37. Skriver, H. Crop classification by multitemporal C- and L-band single- and dual-polarization and fully polarimetric SAR. *IEEE Trans. Geosci. Remote Sens.* **2012**, *50*, 2138–2149. [CrossRef]

38. Lee, J.S.; Grunes, M.R.; Pottier, E. Quantitative comparison of classification capability: Fully polarimetric versus dual and single-polarization SAR. *IEEE Trans. Geosci. Remote Sens.* **2001**, *39*, 2343–2351.

39. Waske, B.; Benediktsson, J.A. Fusion of support vector machines for classification of multisensor data. *IEEE Trans. Geosci. Remote Sens.* **2007**, *45*, 3858–3866. [CrossRef]

40. Lardeux, C.; Frison, P.L.; Tison, C.; Souyris, J.C.; Stoll, B.; Fruneau, B.; Rudant, J.P. Support vector machine for multifrequency SAR polarimetric data classification. *IEEE Trans. Geosci. Remote Sens.* **2009**, *47*, 4143–4152. [CrossRef]
41. Mountrakis, G.; Im, J.; Ogole, C. Support vector machines in remote sensing: A review. *Photogramm. Remote Sens.* **2011**, *66*, 247–259. [CrossRef]
42. Strahler, H. The use of prior probabilities in MaximumLikelihood Classification of remotely sensed data. *Remote Sens. Environ.* **1980**, *10*, 135–163.
43. Huang, C.; DAVIS, L.S.; Townshen, J.R.G. An assessment of support vector machines for land cover classification. *Int. J. Remote Sens.* **2002**, *23*, 725–749. [CrossRef]
44. Wieland, M.; Liu, W.; Yamazaki, F. Learning Change from Synthetic Aperture Radar Images: Performance Evaluation of a Support Vector Machine to Detect Earthquake and Tsunami-Induced Changes. *Remote Sens.* **2016**, *8*, 792. [CrossRef]
45. Vapnik, V. *The Nature of Statistical Learning Theory*, 2nd ed.; Springer: New York, NY, USA, 2000; pp. 1–340.
46. Pal, M.; Mather, P.M. Support vector machines for classification in remote sensing. *Int. J. Remote Sens.* **2005**, *26*, 1007–1011. [CrossRef]
47. Chang, C.C.; Lin, C.J. LIBSVM: A library for support vector machines. *J. ACM Trans. Intell. Syst. Technol.* **2001**, *2*, 1–23. [CrossRef]
48. Hsu, C.-W.; Chang, C.-C.; Lin, C.-J. *A Practical Guide to Support Vector Classification*; Technical Report; Department of Computer Science, National Taiwan University: Taipei, Taiwan, 2010; pp. 1–16.
49. Wu, T.F.; Lin, C.J.; Weng, R.C. Probability estimates for multi-class classification by pairwise coupling. *J. Mach. Learn. Res.* **2004**, *5*, 975–1005.
50. Paneque-Gálvez, J.; Mas, J.F.; Moré, G.; Cristóbal, J.; Orta-Martínez, M.; Luz, A.C.; Reyes-García, V. Enhanced land use/cover classification of heterogeneous tropical landscapes using support vector machines and textural homogeneity. *Int. J. Appl. Earth Obs. Geoinf.* **2013**, *23*, 372–383. [CrossRef]

*applied*
*sciences*

MDPI

*Article*

# Dual-Branch Deep Convolution Neural Network for Polarimetric SAR Image Classification

Fei Gao [1], Teng Huang [1,2], Jun Wang [1,\*], Jinping Sun [1], Amir Hussain [3] and Erfu Yang [4]

[1]    School of Electronic and Information Engineering, Beihang University, Beijing 100191, China;
       feigao2000@163.com (F.G.); huangteng1220@buaa.edu.cn (T.H.); sunjinping@buaa.edu.cn (J.S.)
[2]    Department of Computer and Electronic Engineering, Wuzhou University, Wuzhou 543000, China
[3]    Cognitive Signal-Image and Control Processing Research Laboratory, School of Natural Sciences University
       of Stirling, Stirling FK9 4LA, UK; hussain.doctor@gmail.com
[4]    Space Mechatronic System Technology Laboratory, Department of Design, Manufacture and Engineering
       Management University of Strathclyde, Glasgow G1 1XJ, UK; erfu.yang@strath.ac.uk
\*    Correspondence: wangj203@buaa.edu.cn; Tel.: +86-135-8178-4500

Academic Editors: Carlos López-Martínez and Juan Manuel Lopez-Sanchez
Received: 9 March 2017; Accepted: 24 April 2017; Published: 27 April 2017

**Abstract:** The deep convolution neural network (CNN), which has prominent advantages in feature learning, can learn and extract features from data automatically. Existing polarimetric synthetic aperture radar (PolSAR) image classification methods based on the CNN only consider the polarization information of the image, instead of incorporating the image's spatial information. In this paper, a novel method based on a dual-branch deep convolution neural network (Dual-CNN) is proposed to realize the classification of PolSAR images. The proposed method is built on two deep CNNs: one is used to extract the polarization features from the 6-channel real matrix (6Ch) which is derived from the complex coherency matrix. The other is utilized to extract the spatial features of a Pauli RGB (Red Green Blue) image. These extracted features are first combined into a fully connected layer sharing the polarization and spatial property. Then, the Softmax classifier is employed to classify these features. The experiments are conducted on the Airborne Synthetic Aperture Radar (AIRSAR) data of Flevoland and the results show that the classification accuracy on 14 types of land cover is up to 98.56%. Such results are promising in comparison with other state-of-the-art methods.

**Keywords:** polarimetric SAR images; deep convolution neural network; dual-branch convolution neural network; land cover classification

---

## 1. Introduction

Polarimetric synthetic aperture radar (PolSAR) is a kind of high resolution imaging system, which can work under all weather, day-and-night conditions. The PolSAR data can be used to describe the scattering mechanism of the earth surface and provide rich information for terrain surface classification with the complex coherency matrix [1], scattering matrix [2], etc. With the development of the PolSAR system, PolSAR data such as Advanced Synthetic Aperture Radar (ASAR)/Environmental Satellite (ENVI-SAT), Phased Array L-band Synthetic Aperture Radar (PALSAR)/Advanced Land Observing Satellite (ALOS) and Radar Satllite-2 are becoming more and more available, thus PolSAR image classification has become an important research topic [3].

It is a challenge to automatically extract and select features in PolSAR image classification. Traditional methods generally extract features manually per the scattering characteristics of the terrain surface. The features include radiation information [4], polarization information [5], sub-aperture decomposition [6], decomposition information [7], etc. A single feature or combined features are then fed into an appropriate classifier for classification. Using features individually cannot achieve satisfactory performance. Even if these features are combined then, the classification accuracy will not be improved due to subjectiveness. In addition, the features that can be combined are limited, and the computational complexity will grow with the increase of the amount of combined features. Therefore, traditional methods cannot make full use of the rich features of the PolSAR data to improve the classification accuracy.

In recent years, the theory of deep learning has set off a wave in the field of pattern recognition. In 2006, Hinton et al. proposed an unsupervised greedy method based on Deep Belief Network (DBN), which trains layer by layer, solving the vanishing gradient problem caused by deep training [8]. In 2011, Wong et al. proposed the regularized deep Fisher mapping (RDFM) method per Fisher criterion to enhance the feature separability by using the neural network algorithm, which can eliminate the overfitting problem [9]. Then, many scholars put forward a variety of deep learning models based on different application backgrounds, such as Deep Restricted Boltzmann (DRB) [10], Stacked Denoising Autoencoders (SDA) [11], and the Deep Convolutional Neural Network (CNN) [12]. As a pillar of deep learning, the CNN is one of the best models for solving the "perception" issues. For example, the AlexNet model won first prize in the ImageNet ILSVRC image classification contest in 2012, which has caused widespread concern in related fields [13]. To solve the problem of inefficient utilization of features in PolSAR image processing, some scholars introduce the deep learning framework to extract features of PolSAR images. Wang et al. converted the PolSAR image into the scattering matrix, and then established multichannels for the CNN model [14]. Afterwards, the features were extracted automatically and the images were classified by wide training. Experimental results showed that the PolSAR image classification algorithm based on the CNN is higher than that of the traditional algorithms using the same dataset. Zhou et al. converted the complex matrix of the PolSAR image into a real matrix of six channels to suit the input of the neural network, and designed two, cascaded, fully connected networks to map the features to a certain classifier [15]. This algorithm further improves the accuracy of PolSAR image classification.

Although the deep learning framework provides an idea to solve the problem of low utilization of rich features in PolSAR image classification, it still faces the following problems. First, SAR image is a special kind of microwave image and the regions with the same gray level do not necessarily have similar optical properties. Therefore, the existing methods for optical images based on the deep learning framework may not be suitable for PolSAR image processing. Second, the existing PolSAR image classification methods based on deep learning only consider the polarization features of the image, while ignoring the spatial features.

To solve the above problems, this paper proposes a dual-branch deep convolution neural network (Dual-CNN) method for PolSAR image classification. The proposed method is composed of two CNNs: one is used to extract the polarization features of the real matrix of the six channels (6Ch-CNN) and the other is used to extract the spatial features of the Pauli RGB image (PauliRGB-CNN). These two kinds of features are fed into a fully connected layer to achieve mutual harmony, and then the Softmax classifier is followed immediately to complete the classification work.

The remainder of this paper is structured as follows. Basics of the CNN are introduced in Section 2. In Section 3, we present the architecture of the proposed method. Experiment results and discussions are given in Section 4. In Section 5, the conclusion is made.

## 2. Basics of the CNN

A typical CNN is composed of an input layer, convolution layer, pooling layer and output layer. The input layer receives the pixels from the image. The convolution layer utilizes the convolution kernel to extract image features. The pooling layer is followed by the convolution layer, aiming at reducing the pixels to be processed and formulating the abstract features. The output layer maps the extracted features into classification vectors corresponding to the feature categories. The training of the CNN has two processes: the Forward Propagation and the Backward Propagation.

## 2.1. The Forward Propagation

The Forward Propagation (FP) is a mapping process where the output of the previous layer is taken as the input of the current layer. To avoid the defects of the linear model, neurons of each layer should be added with a nonlinear activation function in the mapping process. Since the first layer only receives pixel values, there are no activation functions. From the second layer to the last layer, nonlinear activation functions are employed. Thus, the output of each layer can be expressed as:

$$\left.\begin{aligned} z^l &= W^l * x^{l-1} + b^l \\ a^l &= \sigma(z^l) \end{aligned}\right\} \tag{1}$$

where $l$ represents the $l^{th}$ layer, and $*$ means convolution operation. $W^l$, $b^l$, and $z^l$ are the weights matrix (for the convolution layer, it is the convolution kernel), the bias matrix and weighted input of the $l^{th}$ layer respectively. $\sigma$ is the nonlinear activation function. If $l = 2$, then $x^{2-1} = x^1$ is the image matrix whose elements are pixel values. If $l > 2$, then $x^{l-1}$ is the feature maps matrix $a^{l-1}$, which is extracted from the $(l-1)^{th}$ layer i.e., $x^{l-1} = a^{l-1} = \sigma(z^{l-1})$. Suppose $L$ is the output layer, $a^L$ represents the final output vector.

## 2.2. The Backward Propagation

The Backward Propagation (BP) algorithm is a supervised learning method. It first selects a cost function based on the output and the targeted values, then calculates the error vectors, and lastly applies the Gradient Descent (GD) to update $W^l$ and $b^l$ parameters. Specifically:

1.  Selection of cost function. The quadratic function is the common cost function. However, it would be time-consuming if the neurons make an obvious mistake during the training process. Alternatively, we take Cross-Entropy ($E_0^L$) as the cost function which is determined by Equation (2):

$$E_0^L = -\frac{1}{n} \sum_{i=1}^{n} \sum_{k=1}^{N} \left[ t_k^L ln a_k^L + (1 - t_k^L) ln(1 - a_k^L) \right] \tag{2}$$

where $n$ is the total number of training sets, and $N$ is the number of neurons in the output Layer corresponding to the $N$ classes. $t_k^L$ is the targeted value corresponding to the $k^{th}$ neuron of the output layer, and $a_k^L$ is the actual output value of the $k^{th}$ neuron of the output layer.

2.  Calculation of error vectors. The error vector of the output layer $L$ is defined by

$$\delta^L = \frac{\partial E_0^L}{\partial z^L} \tag{3}$$

where the symbolic $\partial(\cdot)$ represents the partial derivative operation. Back-propagate the error vector $\delta^L$. For each $l = \{L - 1, L - 2, ..., 2\}$, $\delta^l$ can be computed by the Chain Rule as:

$$\delta^l = W^{l+1}\delta^{l+1} \circ \sigma'(z^l) \tag{4}$$

where the symbolic $\circ$ is the Hadamard product (or Schur product) which denotes the element-wise product of the two vectors.

3.  Updates of weights and the bias matrix. The gradients of $W^l$ and $b^l$ are denoted as $\frac{\partial E_0^L}{\partial W^l}$ and $\frac{\partial E_0^L}{\partial b^l}$ respectively. The partial derivative of $E_0^L$ to $W^l$ and $b^l$ can be calculated with Equations (1) and (3):

$$\left.\begin{aligned} \frac{\partial E_0^L}{\partial W^l} &= \frac{\partial E_0^L}{\partial a^l} \circ \frac{\partial a^l}{\partial W^l} = \delta^l \circ x^{l-1} \\ \frac{\partial E_0^L}{\partial b^l} &= \frac{\partial E_0^L}{\partial a^l} \circ \frac{\partial a^l}{\partial b^l} = \delta^l \end{aligned}\right\} \tag{5}$$

The change values of $W^l$ and $b^l$: $\Delta W^l$ and $\Delta b^l$, can be calculated respectively by

$$\left.\begin{aligned} \Delta W^l &= -\eta \frac{\partial E_0^L}{\partial W^l} \\ \Delta b^l &= -\eta \frac{\partial E_0^L}{\partial b^l} \end{aligned}\right\} \tag{6}$$

where $\eta$ represents the learning rate.

### 2.3. Feature Extraction

In the training process, the CNN is used to extract the features from the data based on the convolution operation. The convolution operation includes a convolution layer followed immediately by a pooling layer. The feature extraction process is shown in Figure 1. From bottom to top, $a$ represents the input data, $b$ represents the feature data which is obtained by the convolution operation and the ReLu (Rectified Linear Units) activation function, and $c$ represents the feature data which is obtained after the pooling process. Red and green patches represent different salient features, blue patches represent the label of the data(or interesting target), the purple patch in $a$ represents the convolution kernel, and the purple patch in $b$ represents the feature which is obtained from the purple patch in $a$ by the convolution operation and the ReLu activation function. More specifically, the CNN works as follows.

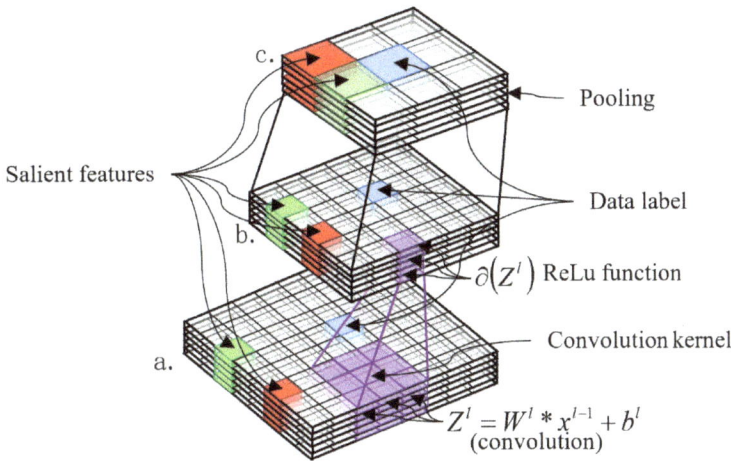

**Figure 1.** The processing of feature extraction in deep convolution neural network (CNN).

First, the salient features of $a$ are preserved, and then passed to a ReLu function for post processing. Second, the non-salient features of $a$ are filtered out by ReLu via setting the minus in the feature map to 0. The derived features are $b$. Finally, more abstract features $c$ of the input data will be obtained based on the pooling layer. If features $c$ are not abstract enough, a second convolution operation is needed. The process is repeated until the most representative features are obtained. This results in deepening the layers of the CNN. In the whole process, the data-label of $a$ may be filtered out, but its main neighborhood features can be preserved for judging $a$-label. In addition, it is worth noting that there is no clear conclusion how many layers of convolution operations are appropriate. Thus, the visual convolution operation is needed.

## 3. The Proposed Method

The proposed method consists of two frameworks: PolSAR data pre-processing and Dual-CNN model design. As shown in Figure 2, the Dual-CNN includes the 6Ch-CNN and PauliRGB-CNN. The polarization features and the spatial features are generated from the pre-processed data by the 6Ch-CNN and PauliRGB-CNN; then, these features are combined by a fully connected layer. In this paper, the above combined features are named as P-S features. Finally, the P-S features are classified by the Softmax classifiers.

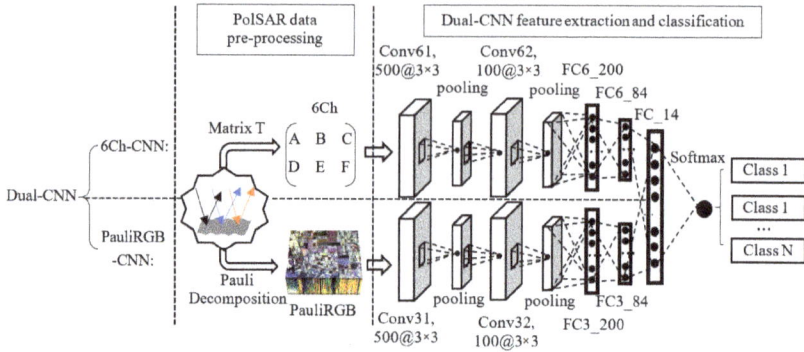

**Figure 2.** The main procedures of the polarimetric synthetic aperture radar (PolSAR) images classification based on the dual-branch deep convolution neural network (Dual-CNN) model.

### 3.1. PolSAR Data Pre-Processing

Since the obtained PolSAR data contain the complex coherent matrix, they cannot be directly fed into the Dual-CNN model and a pre-processing is required. The pre-processing of the PolSAR data contains three steps: creating a 6Ch to allow the input and representation of polarimetric data; generating a Pauli RGB image to obtain the spatial feature; patching the images with fixed size to adapt to the CNN.

### 3.1.1. Creating 6Ch to Represent the Polarimetric Data

Under the multi-look and reciprocity assumption, the single station PolSAR can be represented by the $3 \times 3$ complex coherent matrix T which is symmetrical. To adapt the input format of the convolution neural network, it is necessary to convert the data into a real matrix. We create a 6Ch to represent the polarimetric data, and each channel is obtained by Equation (7):

$$\left.\begin{array}{l} A = 10ln(SPAN) \\ B = T_{22}/SPAN \\ C = T_{33}/SPAN \\ D = |T_{12}|/\sqrt{T_{11}T_{22}} \\ E = |T_{13}|/\sqrt{T_{11}T_{33}} \\ F = |T_{23}|/\sqrt{T_{11}T_{33}} \end{array}\right\} \tag{7}$$

where $T_{11}$, $T_{22}$, $T_{33}$ represent the diagonal elements of the matrix **T** and they are real numbers while $T_{12}$, $T_{13}$, $T_{23}$ represent complex elements. $A$ is the total scattering power in decibels, here $SPAN = T_{11} + T_{22} + T_{33}$ ; $B$ and $C$ are normalized power of $T_{22}$ and $T_{33}$; $D$, $E$ and $F$ are the relative correlation coefficients. Except $A$, the remaining five parameters are normalized to [0, 1].

Thus, the PolSAR data are converted into a 6Ch to form a $6 \times m \times n$ dataset, where 6 represents the total number of the channels, i.e., *A, B, C, D, E* and *F*; *m* and *n* represent the number of rows and columns in a single channel, respectively.

### 3.1.2. Generating Pauli RGB Image to Obtain the Spatial Feature

The Pauli decomposition of the scattering matrix *S* is often employed to represent all the polarimetric information in a single PolSAR image, and its form is:

$$S = \begin{bmatrix} S_{HH} & S_{HV} \\ S_{VH} & S_{VV} \end{bmatrix} = aS_a + bS_b + cS_c + dS_d \tag{8}$$

where $S_a$, $S_b$, $S_c$ and $S_d$ constitute a set of orthogonal Pauli bases, and *a, b, c* and *d* are coefficients. $S_a$ is the odd scattering mechanism, representing the terrain scattering body; $S_b$ is the dihedral scattering mechanism rotating $0°$ around the axis, and its echo polarization and incident polarization are on the mirror symmetry; $S_c$ is the dihedral scattering mechanism rotating $45°$ around the axis, and its echo polarization and incident polarization are orthogonal; $S_d$ is the antisymmetric component. Since the corresponding scattering mechanism does not exist in the nature, the weighted coefficient *d* is 0 generally. After the Pauli decomposition, the Pauli RGB image is synthetized by a pseudo-color process using the energy corresponding to *a, b* and *c*, as is described in Equation (9):

$$|a|^2 \rightarrow Blue, |b|^2 \rightarrow Red, |c|^2 \rightarrow Green \tag{9}$$

The synthesised Pauli RGB image contains rich contour, texture, and color features, which is in greet agreement with those of real ground scenes. This enables recognition by the naked eye. In addition, the Pauli RGB image can reduce the interference of other data on the feature extraction, improving algorithm robustness. Furthermore, the CNN is good at dealing with color images, so the Pauli RGB image is suitable. For these reasons, many classification algorithms use the Pauli RGB image as their input [16,17], and so does our method.

### 3.1.3. Patching the Images with Fixed Size

It is required that the CNN processes the data with a fixed size. However, different targets usually have different sizes, so it is difficult to use a generalized size for all target slices. Thus, some researchers propose to process the patches via stretching or filling the bounding of the image with 0 pixels. Although these methods can solve this problem to a certain extent, they will bring some unexpected errors. For example, if the boundary of small objects is filled with 0, the detection accuracy of the targets in complex environments will be limited, which thereby influences its feature learning. The details of our patching method are as follows.

First, based on the sliding window with a fixed size, we traverse the entire dataset to obtain a set of fixed size slices for each channel of the 6Ch and Pauli RGB image, which is shown in Figure 3.

By this way, the maximum number of slices for each channel can be obtained by:

$$n = (w - s + i)(h - s + i)/i^2 \tag{10}$$

where *w* and *h* are the width and height of each channel data respectively, *s* is the size of the sliding window, and *i* is the span while sliding. It should be noted that *s* and *i* are relevant to the size of the target. If the target is small (e.g., several pixels), then *i* is minor and *s* should be carefully selected. The selection of both parameters will be detailed in the experiment section.

Second, we assign a label for each slice. Specifically, each pixel in ground-truth is assigned a label per the category it belongs to, and then we choose the location of the center pixel of the slice as the index to search the category label in the ground-truth. The selected label is finally assigned for the slice.

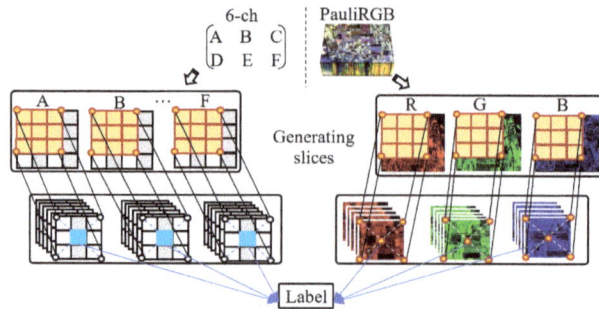

**Figure 3.** The generation of the fixed-size slices based on the center pixel.

In this manner, all samples are obtained with labels attached, and we can ensure that the CNN can process the small target in the complex environment.

Using a sliding window can get enough training sets, but the samples will be more or less repeated. Therefore, before feeding the data into the CNN, we need to reduce the data redundancy by principle component analysis (PCA).

### 3.2. Feature Extraction and Classification Based on the Dual-CNN Model

The Dual-CNN model consists of two CNNs, i.e., the 6Ch-CNN and PauliRGB-CNN. The 6Ch-CNN contains two convolution layers (Conv61, Conv62), two pooling layers, and two fully connected layers (FC6_200, FC6_84). It is used to acquire the polarization data feature. The PauliRGB-CNN also includes two convolution layers (Conv31, Conv32), two pooling layers and two fully connected layers (FC3_200, FC3_84), and it is applied to obtain the spatial features. More specifically, as shown in Figure 2, "Conv61, 500@3×3" represents the first 6-channel convolution layer depending upon the 3×3 convolution kernel and generates 500 feature maps. "FC6_200" represents the 6-channel fully connected layer consisting of 200 neurons. Notice that the ReLu is used as the activation function for all the hidden layers and the 2×2 max-pooling is used for the pooling layers. PauliRGB-CNN is constructed as done in the 6Ch-CNN.

In the training process, the FP and BP are two vital procedures for updating the network. By training the network, the 6Ch-CNN can obtain the features with the property of polarization, while PauliRGB-CNN obtains the features which contain spatial characteristics. Then, the Softmax function is employed to implement the classification.

#### 3.2.1. The Forward Propagation of the Dual-CNN Model

In the 6Ch-CNN, the input polarization data are a 6Ch whose size is fixed. The polarization feature $F_1(p_n)$ can be obtained by Equation (1) and pooling. For the PauliRGB-CNN, the spatial feature $F_2(s_n)$ can be obtained in the same manner, and the input Pauli RGB image is a fixed size slice with three channels. Next, two kinds of data are input into the 6Ch-CNN and PauliRGB-CNN separately to obtain the respective features. Then, the obtained two kinds of features are fed into a fully connected layer to combine with each other, and the P-S features $F^{(n)}$ can be represented as:

$$F^{(n)} = \partial[W_1 * (F_1(p_n) \odot F_2(s_n)) + b_1] \tag{11}$$

where $W_1$ and $b_1$ represent the weights and bias matrix in the last fully connected layer, and the joint operator $\odot$ stacking the former and the latter items to be input of the last fully connected layer. At last, $F^{(n)}$ is put into the Softmax to produce a probability vector for each class:

$$P^{(n)} = \frac{1}{\sum_{k=1}^{N} e^{W_k F^{(n)} + b_k}} \begin{bmatrix} e^{W_1 F^{(n)} + b_1} \\ e^{W_2 F^{(n)} + b_2} \\ \vdots \\ e^{W_N F^{(n)} + b_N} \end{bmatrix} \tag{12}$$

where $N$ is the total number of the classes. $P_{max}^{(n)}$ is the max probability in the $N$-dimensional vector $P^{(n)}$, and it is recognized as the predicted result.

### 3.2.2. The Backward Propagation of the Dual-CNN Model

The cost function is established using the ground-truth after obtaining the output category through the FP. In our method, the Cross Entropy is selected as the cost function. In addition, the weights and bias can be obtained on the given training set according to Equations (2)–(4), and (6) by the BP process.

In order to improve the performance of BP, the Adam optimization algorithm is used in the process of batching gradient descent. The weights in each layer are initialized by a group of values which are subject to the Gaussian random distribution in a certain interval given in Equation (13):

$$\left[ -4\sqrt{\frac{6}{f_{in} + f_{out}}}, 4\sqrt{\frac{6}{f_{in} + f_{out}}} \right] \tag{13}$$

where $f_{in}$ and $f_{out}$ are the numbers of the input and the output feature maps at each layer respectively.

## 4. Experiment

To verify the performance of the proposed Dual-CNN model, we conduct an experiment on the Flevoland full PolSAR data. The experiment contains the following three aspects:

1. Comparing our method with the single-branch network, i.e., the 6Ch-CNN and PauliRGB-CNN model.
2. Comparing our method with some classical algorithms and some recently proposed classification algorithms with the same dataset.
3. Discussing how the size of the slices influences the performance of our method, and then conducting research on the visual representation of the features.

### 4.1. Flevoland Data

Flevoland full PolSAR data are farmland images at L wave band and were acquired by the AIRSAR Aircraft platform in 16 August 1989. It contains *HH, HV, VH* and *VV* (H and V represent horizontal polarization and vertical polarization respectively) channels of polarimetric information. Each channel has $750 \times 1024$ pixels. The resolution of the image is 6.6 m in the range direction and 12.1 m in the azimuth direction. Since the complex coherent matrix $T$ can describe the scattering mechanism, we transform the 4-channel original data to $T$. According to Section 3.1, we convert the $T$ in the 6Ch and 4-channel original data to the Pauli RGB image. Figure 4a depicts a Pauli RGB image which includes crops, lake and lands, etc. In this experiment, we first choose 14 types of land cover classes to complete the classification. Then we use the ArcGIS to obtain the ground-truth image of Flevoland according to the Pauli RGB image and google earth. The ground-truth image of Flevoland is shown in Figure 4b.

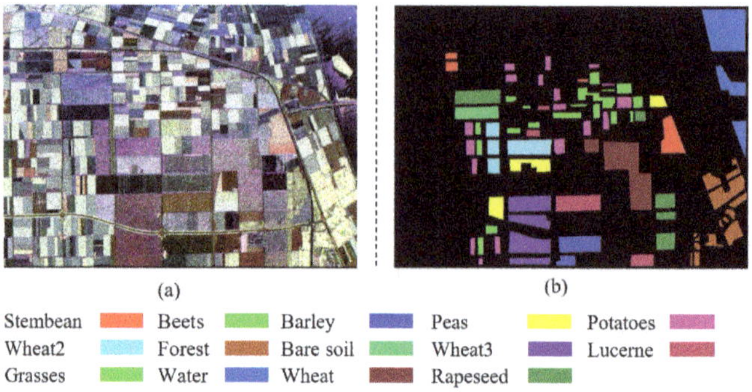


**Figure 4.** (**a**) Pauli RGB image of Flevoland PolSAR data; (**b**) Ground-truth image of Flevoland.

When we obtain the 6Ch and Pauli RGB image, we can acquire slices as illustrated in Section 3.1.3. In Equation (10), we set $s$ to 15 and $i$ to 5. Then, we obtain $(750 - 15 + 5) \times (1024 - 15 + 5)/5^2$ slices with the size of $15 \times 15$. Due to the fact that the images are converted into the 6Ch and 3-channel Pauli RGB image, the numbers of the slices of two branches are $6 \times (750 - 15 + 5) \times (1024 - 15 + 5)/5^2$ and $3 \times (750 - 15 + 5) \times (1024 - 15 + 5)/5^2$ respectively. Subsequently, we label the slices per ground-truth. Finally, we divide the slices into two parts, one is the training set and the other is the testing set. Usually, we chose 75% slices as the training set and the remaining slices as the testing set. It is worth noting that the slices in the same location of the 6Ch and Pauli RGB image should be assigned to an individual part. Table 1 shows the terrain training set and the testing set in the 6Ch and Pauli RGB image; it only includes 14 types of land cover on the ground-truth image.

**Table 1.** The detailed information of the training set and testing set on 14 types of land cover classes on Flevoland PolSAR data.

| Label | Type | Color | Train | | Test | |
|---|---|---|---|---|---|---|
| | | | **6Ch** | **PauliRGB** | **6Ch** | **PauliRGB** |
| 1 | Stembeans | | 5082 | 5082 | 1693 | 1693 |
| 2 | Beets | | 6039 | 6039 | 2012 | 2012 |
| 3 | Barley | | 5106 | 5106 | 1701 | 1701 |
| 4 | Peas | | 5530 | 5530 | 1843 | 1843 |
| 5 | Potatoes | | 9180 | 9180 | 3060 | 3060 |
| 6 | Wheat2 | | 7343 | 7343 | 2447 | 2447 |
| 7 | Forest | | 10,093 | 10,093 | 3364 | 3364 |
| 8 | Bare soil | | 3299 | 3299 | 4099 | 4099 |
| 9 | Wheat3 | | 12,663 | 12,663 | 4221 | 4221 |
| 10 | Lucerne | | 6872 | 6872 | 2290 | 2290 |
| 11 | Grasses | | 4200 | 4200 | 1399 | 1399 |
| 12 | Water | | 14,739 | 14,739 | 4913 | 4913 |
| 13 | Wheat | | 12,361 | 12,361 | 4120 | 4120 |
| 14 | Rapeseed | | 9013 | 9013 | 2838 | 2838 |
| Total | - | - | 111,520 | 111,520 | 37,000 | 37,000 |

## *4.2. Comparing with One-CNN*

To verify the effectiveness of the Dual-CNN model, we compare the proposed method with the 6Ch-CNN and PauliRGB-CNN. We apply a Softmax classifier after their own respective last layers. In this way, we can train and test the networks with the same parameters.

The training process of our method is performed iteratively 100 times on NVIDIA's GeForce GTX 1070 with 8GB of GPU memory, and 1000 training samples are used in every epoch. Figures 5a, 6a and 7a show the loss curve of the 6Ch-CNN, PauliRGB-CNN and Dual-CNN, where the horizontal axis represents the number of epochs and the vertical axis denotes the loss value. Figures 5b, 6b and 7b show the accuracy curve of the 6Ch-CNN, PauliRGB-CNN and Dual-CNN, where the horizontal axis represents the number of epochs and the vertical axis denotes the classification accuracy. In addition, the blue line depicts the training curve and the green line indicates the testing curve.

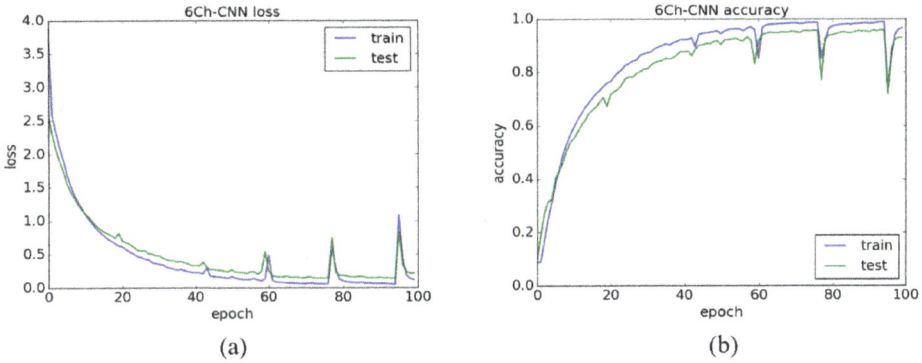

**Figure 5.** (**a**) Loss curve of the 6Ch-CNN; (**b**) accuracy curve of the 6Ch-CNN.

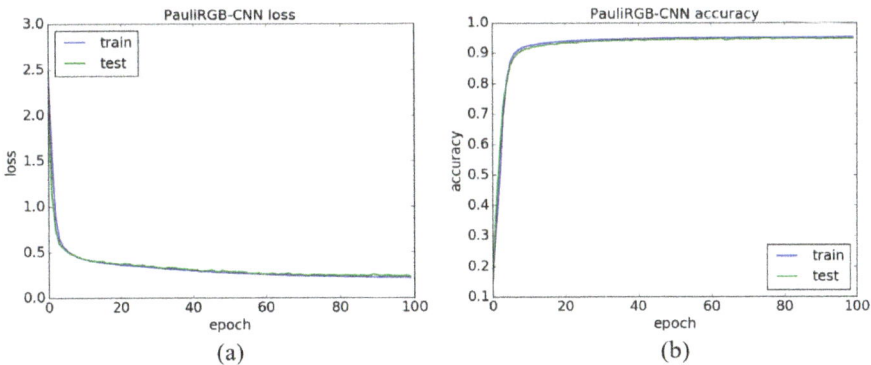

**Figure 6.** (**a**) Loss curve of the PauliRGB-CNN; (**b**) accuracy curve of the PauliRGB-CNN.

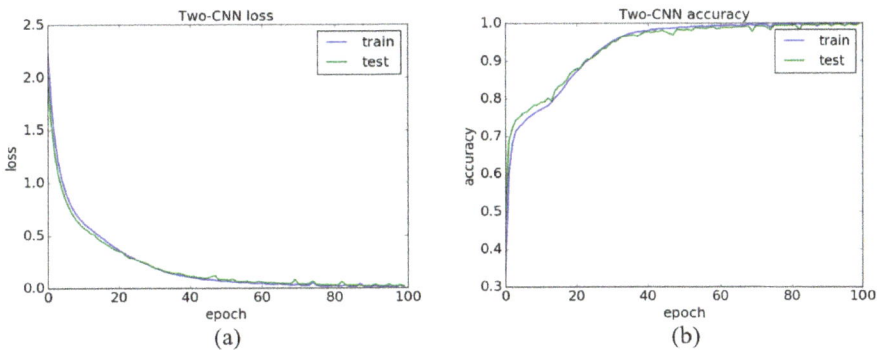

**Figure 7.** (**a**) Loss curve of the Dual-CNN; (**b**) accuracy curve of the Dual-CNN.

As shown in Figure 5, the loss curve of the 6Ch-CNN is not very stable, and its accuracy curve has a large fluctuation as shown in Figure 5b. It is due to the fact that the 6Ch-CNN cannot obtain the general features of the same class in complicated polarimetric data. Moreover, compared with the 6Ch-CNN, although the loss curve and accuracy curve are stable in PauliRGB-CNN, the 3-channel data of the Pauli RGB image are scarce. This fact causes the accuracy rate of the PauliRGB-CNN to be below 95%.

However, in Figure 7b, it can be found that the training accuracy of the Dual-CNN reaches 100%. The testing accuracy is becoming coincident, keeping at 98%. The training loss and the testing loss are stable except for the 45th, 70th and 75th epoch. It proves that the polarimetric features and spatial features are well combined. In Figure 7a, the loss value of the Dual-CNN model decreases, beginning with a minor value 2.25, in which case the training accuracy is 50%, cutting down the training time and the number of epochs. That proved the validity of Equation (13). We draw the conclusion in terms of the accuracy that although there are some anomalies in the training set, the Dual-CNN model is not impacted.

To clarify the result, we list the accuracy rate of the three different methods in Table 2. For the classification of 14 types of land cover classes, the lowest accuracy rate of the Dual-CNN model is still above 95%. Especially, the accuracies of Wheat2 (Label: 6) and Bare soil (Label: 8) reach 100%. However, the average accuracies of the 6Ch-CNN and PauliRGB-CNN are 5.71% and 4.45% lower than the Dual-CNN model.

**Table 2.** The detailed classification accuracy of the Dual-CNN, 6Ch-CNN, and PauliRGB-CNN on Flevoland PolSAR data.

| Label | Dual-CNN (%) | 6Ch-CNN (%) | PauliRGB-CNN (%) |
|---|---|---|---|
| 1 | 97.77 | 96.04 | 95.64 |
| 2 | 98.21 | 90.85 | 90.70 |
| 3 | 97.88 | 93.94 | 94.17 |
| 4 | 96.72 | 91.91 | 93.67 |
| 5 | 95.96 | 88.56 | 92.57 |
| 6 | 100 | 95.05 | 94.26 |
| 7 | 99.94 | 97.08 | 95.97 |
| 8 | 100 | 95.54 | 93.45 |
| 9 | 95.95 | 87.84 | 90.48 |
| 10 | 99.51 | 92.70 | 94.07 |
| 11 | 98.85 | 95.40 | 95.42 |
| 12 | 99.92 | 91.34 | 96.74 |
| 13 | 99.85 | 93.20 | 93.48 |
| 14 | 99.39 | 90.45 | 95.53 |
| overall | 98.56 | 92.85 | 94.01 |

For the convenience of comparison, the classified results are labelled using the same color as the ground-truth. The results of the ground-truth, Dual-CNN, 6Ch-CNN and PauliRGB-CNN are shown in Figure 8a–d, respectively.

Analyzing Figure 8c,d, we find that the 6Ch-CNN and PauliRGB-CNN have obvious faults. The 6Ch-CNN tends to have more scatter errors while PauliRGB-CNN tends to have block errors. However, the Dual-CNN only has a few scatter errors and almost no block errors, which is better than the single branch way. It also demonstrates that the combination of the 6Ch-CNN and PauliRGB-CNN with a fully connected layer is effective for classification.

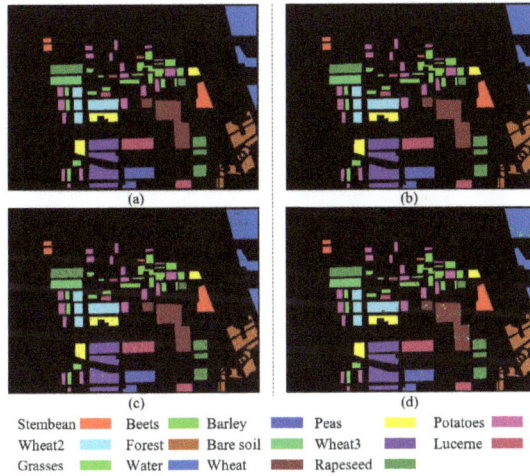

Stembean ▮  Beets ▮  Barley ▮  Peas ▮  Potatoes ▮
Wheat2 ▮  Forest ▮  Bare soil ▮  Wheat3 ▮  Lucerne ▮
Grasses ▮  Water ▮  Wheat ▮  Rapeseed ▮

**Figure 8.** (**a–d**) represent the classification results of the ground-truth, Dual-CNN, 6Ch-CNN, and PauliRGB-CNN, respectively.

### 4.3. Comparing with Other Methods

In this section, we compare our method with some classical algorithms and some newly published methods on the same dataset. The classical algorithms include Maximum Likelihood [18], Support Vector Machine (SVM) [19], and Minimum Distance [20], which are all performed on the ENVI remote sensing image processing platform [21,22]. Lee et al. [23], Zhou et al. [15] and Wang et al. [24] are newly published methods; the method of Lee et al. is an unsupervised algorithm, and those of Zhou et al. and Wang et al. are supervised algorithms.

Table 3 shows the results of the classification. We performed our experiment on both 11 and 14 types of land cover classes. For classical algorithms, SVM has the highest accuracy, but it is still lower than our method. For these newly published methods, supervised algorithms are better than unsupervised ones; the method of Zhou et al. is better than others' supervised algorithms for 11 classes because of using the CNN. However, our method is still the best. In addition, we find that the accuracy decreases when the number of classes increases.

**Table 3.** Comparison of results with other methods.

| Types | Names | Number of Classes | Accuracy (%) |
|---|---|---|---|
| Classical | Maximum Likelihood | 14 | 64.26 |
| | SVM | 14 | 71.29 |
| | Minimum Distance | 14 | 54.66 |
| Newly Published | Lee et al. [23] | 11 | 81.63 |
| | Zhou et al. [15] | 11 | 93.38 |
| | Wang et al. [24] | 11 | 93.24 |
| Proposed | Dual-CNN | 14 | 98.56 |
| | | 11 | 98.93 |

### 4.4. Different Fixed Size Slices and Visualization of Feature Maps

4.4.1. The Effect of Slicing Size on Classification Accuracy

The category of the center pixel in the slice served as the label. In order to evaluate how the slice size influences the performance of the algorithm, we performed the experiment with slices of 11 × 11, 15 × 15 and 19 × 19, which are subject to the span equalling to 5. The results of the classification are shown in Figures 9 and 10.

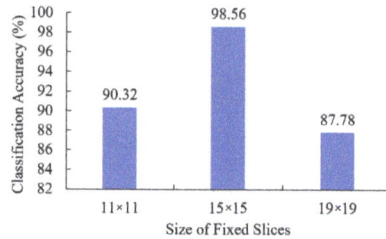

**Figure 9.** The classification accuracy of the Dual-CNN with the slices of different sizes.

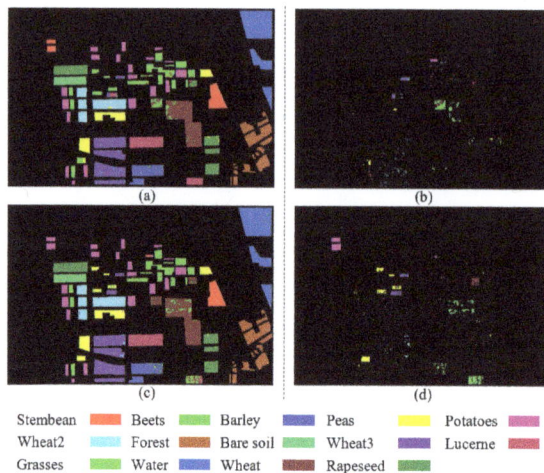

**Figure 10.** (**a**,**c**) represent the classification results of the Dual-CNN with slices of 11 × 11 and 19 × 19 respectively; (**b**,**d**) display the false results in (**a**,**c**).

As shown in Figure 9, the experiment which uses the slices of 11 × 11 and 19 × 19 has lower classification accuracy than that of using slices of 15 × 15. Therefore, the slices of 15 × 15 are appropriate for Flevoland full PolSAR data. As is shown in Figure 10a, for the slices of 11 × 11, the Dual-CNN does not perform well for large area targets such as wheat (Wheat: brown; Wheat3: purple); and as is shown in Figure 10b, for the slices of 19 × 19, the Dual-CNN does not perform well for small area targets such as stem bean, peas and so on. However, for the slices of 15 × 15, as shown in Figure 8b, the Dual-CNN conducted on large or small area targets such as wheat or stem beans and peas is better than that of the other two methods. The slice size affects the Dual-CNN classification accuracy.

If the size of the slice is too small, then the Dual-CNN learns inadequate feature information. Slices of large area targets, such as the wheat category, will contain the reduplicated information. Therefore, the features of large area targets learned by the Dual-CNN are not only small in number but also single, which leads to the low classification accuracy. As the size of slices is enlarged, the learning ability of the feature enhances, and the classification accuracy is improved. However, if the size of slices is too large, it will contain the extra features of other objects, which would cause the features of the small area targets to be submerged by other surrounding features. Thus, the Dual-CNN can extract many useless features of small area targets, and the accuracy of classification will decrease.

#### 4.4.2. The Visualization of Feature Maps

To better represent the Dual-CNN model, the 6Ch-CNN and the PauliRGB-CNN are visualized in this paper. As is shown in Figure 11, the visualization processes of two branches are located at the convolution layer and max-pooling layer. The visualization contents include input data visualization, feature extraction visualization, and convolution kernel visualization.

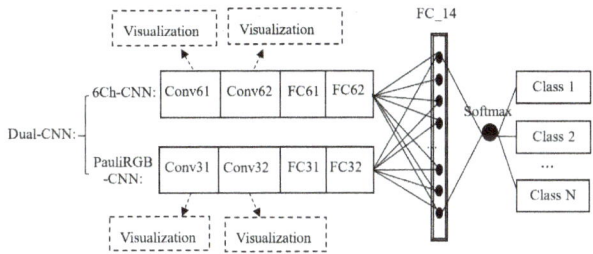

**Figure 11.** The location of the visualization process.

For input data visualization, two Lucerne slices of $15 \times 15$, which come from the 6Ch and Pauli RGB image in the same location, are selected as the samples to show the visualization process. Two slices are named as the 6Ch-input and PauliRGB-input respectively. Figures 12–14 represent the visualization process of the 6Ch-CNN, and Figures 15–17 depict the visualization process of the PauliRGB-CNN. Figures 12 and 15 show the visualized images of 6Ch-input and PauliRGB-input; where Figures 12a and 15a denote the mixed visualized images of the 6Ch-input and PaulRGB-input respectively, Figures 12b and 15b denote their unfolded visualized images in an individual channel. Since the data information of the 6Ch-input and PauliRGB-input are significantly different, it is difficult to infer whether they represent the same object from the visualized images. As shown in Figure 12, both the mixed and unfolded visualized images have scatter pixels for the 6Ch-input. This reflects the various scattering phenomena of the polarized waves. In Figure 15, for the PauliRGB-input, some contours can be observed from the mixed and unfolded visualized images. This indicates the spatial characteristics of the terrain surface. These are the most salient features that the CNN requires, which is beneficial for enhancing the classification accuracy. After this visualization process, the 6Ch-input and the PauliRGB-input are put into the trained Dual-CNN model.

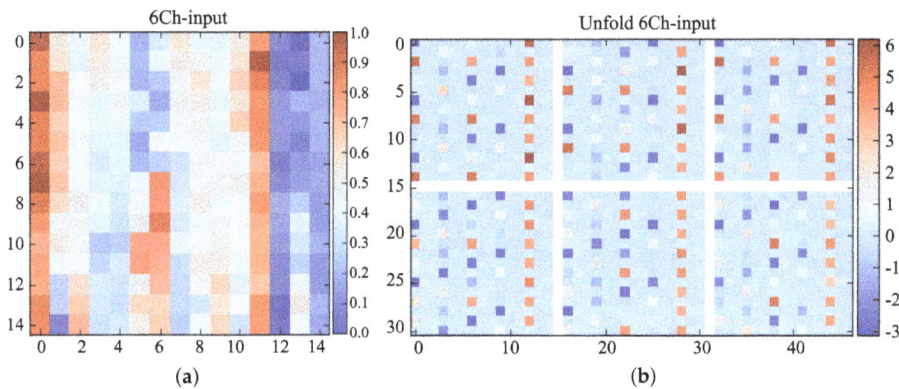

**Figure 12.** (**a**) Mixed visualized image of the 6Ch-input; (**b**) Unfolded visualized images of the 6Ch-input.

As shown in Figure 1, the feature extraction process is performed with three operations, i.e., convoluting, ReLu processing, and pooling processing. Figure 13 depicts the visualized image of the 6Ch-input during extracting features where Figure 13a–c denote the visualization of the convolution operation, ReLu operation and max-pooling operation of the first round of feature extraction; and Figure 13d–f represent the visualization of the second round of feature extraction. As is shown in Figure 13a, the 15 × 15 6Ch-input serves as the input of the 6Ch-CNN. Then, these data are processed by the 3 × 3 convolution kernel in the first layer and are converted into a 14 × 14 feature map as the output. Figure 13 shows the derived 12 visualized images. Compared with Figure 12b, the salient polarization features are preserved since 0 elements (light orange) in the 6Ch-con1 are increased. Figure 13b illustrates the feature map of the 6Ch-con1-reLu1 after ReLu operation; thus, most of the data are set to 0 (blue). By the max-pooling operation, the visualized feature map of 6Ch-con1-reLu1-maxpooling1 after the first round of feature extraction can be obtained, as shown in Figure 13c. Although the polarization features are salient, there exists a lot of redundancy, so a second round of feature extraction is recommended. The processes are shown in Figure 13d–f. Figure 13f shows the final visualized feature map, where the red squares and the orange squares are the basic features of the input slices.

The visualized feature maps of the convolution kernels in the two rounds of feature extraction are shown in Figure 14a,b. From the colorful block in Figure 14, it can be found that the elements of each convolution kernel are not all zero, which indicates that the Dual-CNN has been well trained and the obtained features are obvious.

Figures 16 and 17 show the visualization of feature extraction and convolution kernels for the PauliRGB-input. It illustrates that the PauliRGB-input requires a second round of feature extraction to extract more abstract spatial features as done in the 6Ch-input.

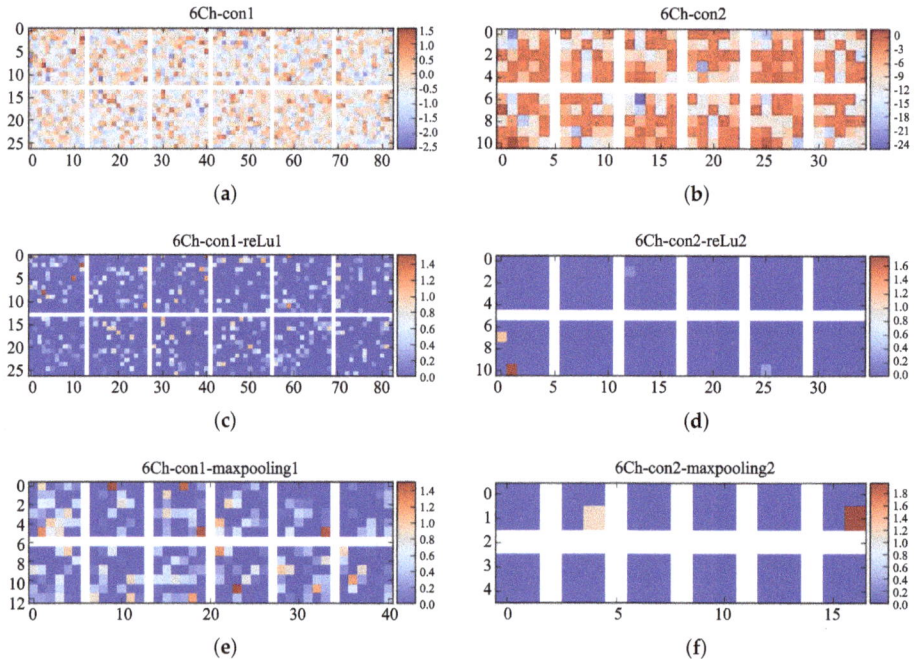

**Figure 13.** Visualized feature maps of the 6Ch-input. (**a**,**c**,**e**) denote the visualized feature maps of the convolution operation, ReLu operation and max-pooling operation in the first round of feature extraction; and (**b**,**d**,**f**) denote the visualization of the second round of feature extraction.

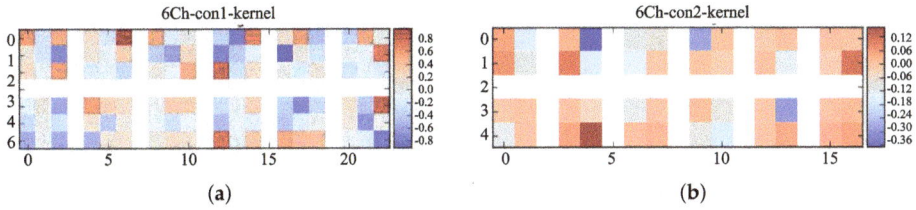

**Figure 14.** Visualization of the convolution kernel of the 6Ch-CNN: (**a**) visualization of the convolution kernel in the first round of feature extraction; and (**b**) visualization of the convolution kernel in the second round of feature extraction.

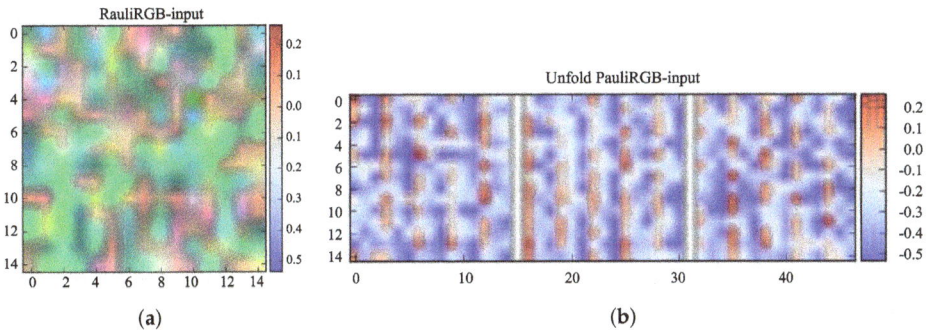

**Figure 15.** (**a**) Mixed visualized image of the PauliRGB-input; (**b**) Unfolded visualized images of the PauliRGB-input.

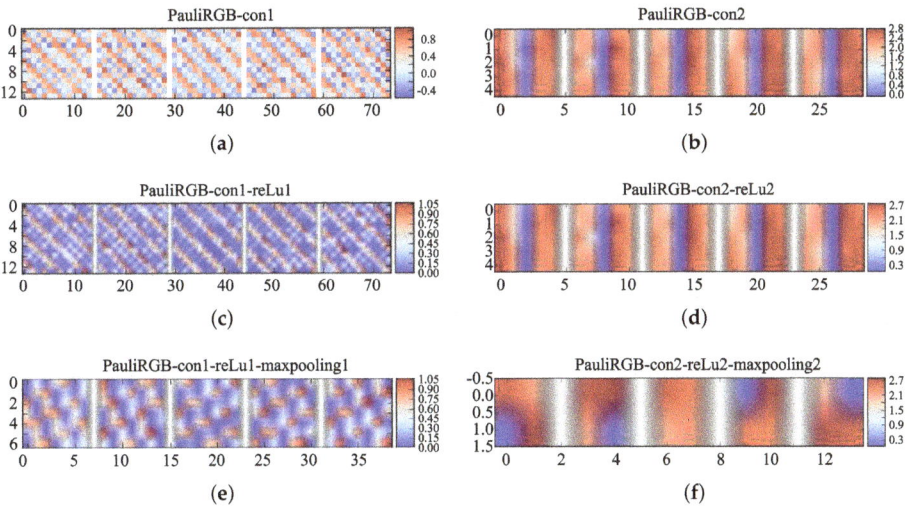

**Figure 16.** Visualized feature maps of the PauliRGB-input. (**a,c,e**) denote the visualized feature maps of the convolution operation, ReLu operation and max-pooling operation in the first round of feature extraction; (**b,d,f**) denote the visualization in the second round of feature extraction.

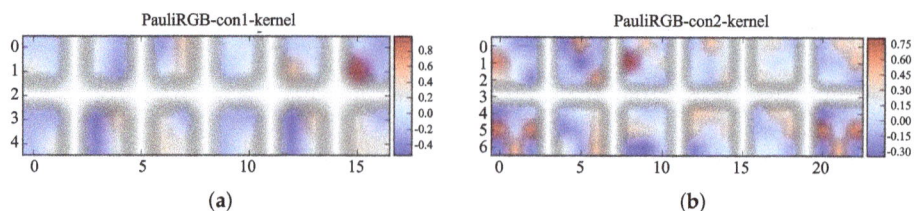

**Figure 17.** Visualization of the convolution kernel of the PauliRGB-CNN: (**a**) visualization of the convolution kernel in the first round of feature extraction; and (**b**) visualization of the convolution kernel in the second round of feature extraction.

Note that the features extracted by the 6Ch-CNN and PauliRGB-CNN have polarimetric and spatial characteristics. Visualization can help to check whether the Dual-CNN model is well trained and to illustrate how to extract the P-S features.

## 5. Conclusions

By exploring the unique characteristics of the PolSAR data, we have presented a new method that achieves excellent accuracy in PolSAR classification. The main contributions of this work lie in the following three aspects. First, we proposed a method of pre-processing the PolSAR data to facilitate subsequent work. Second, a novel CNN framework which consists of two CNNs was presented to extract and fuse the polarization feature and spatial feature of the pre-processed data. Last but not least, visualization of the CNN was applied to help us tune the parameters of the model. We carried out the experiments on 14 types of land cover classes, and the results show that our model is superior to the classical classification methods such as SVM, and Maximum Likelihood. Compared with a single CNN, our method still has higher accuracy due to its P-S features.

**Acknowledgments:** This work was supported by the National Natural Science Foundation of China (61071139; 61471019; 61671035), the Aeronautical Science Foundation of China (20142051022), the Pilot Project (9140A07040515HK01009), the Scientific Research Foundation of Guangxi Education Department (KY 2015LX443), and the Scientific Research and Technology Development Project of Wuzhou City, GuangXi, China (201402205).

**Author Contributions:** Fei Gao and Teng Huang conceived of and designed the experiments. Jun Wang performed the experiments. Jinping Sun, Amir Hussain, Erfu Yang analyzed the data and wrote the paper.

**Conflicts of Interest:** The authors declare no conflict of interest.

## References

1.  Gaber, A.; Soliman, F.; Koch, M.; El-Baz, F. Using full-polarimetric SAR data to characterize the surface sediments in desert areas: A case study in El-Gallaba Plain, Egypt. *Remote Sens. Environ.* **2015**, *162*, 11–28.
2.  Canty, M.J. Images, Arrays, and Matrices. In *Image Analysis, Classification and Change Detection in Remote Sensing: With Algorithms for ENVI/IDL and Python*, 3rd ed.; CRC Press: Boca Raton, FL, USA, 2014; pp. 1–32.
3.  Rosenqvist, A.; Shimada, M.; Ito, N.; Watanabe, M. ALOS PALSAR: A Pathfinder mission for global-scale monitoring of the environment. *IEEE Trans. Geosci. Remote Sens.* **2007**, *45*, 3307–3316.
4.  Yang, M.; Zhang, G. A novel ship detection method for SAR images based on nonlinear diffusion filtering and Gaussian curvature. *Remote Sens. Lett.* **2016**, *7*, 211–219.
5.  Niu, C.; Zhang, G; Zhu, J; Liu, S; Ma, D. Correlation Coefficients Between Polarization Signatures for Evaluating Polarimetric Information Preservation. *IEEE Geosci. Remote Sens. Lett.* **2011**, *8*, 1016–1020.
6.  Deng, L.; Yan, Y.N.; Sun, C. Use of Sub-Aperture Decomposition for Supervised PolSAR Classification in Urban Area. *Remote Sens.* **2011**, *7*, 1380–1396.
7.  Xiang, D.L.; Tang, T.; Hu, C.B.; Fan, Q.H.; Su, Y. Built-up Area Extraction from PolSAR Imagery with Model-Based Decomposition and Polarimetric Coherence. *Remote Sens.* **2016**, *8*, 685.
8.  Hinton, G.E.; Salakhutdinov, R.R. Reducing the dimensionality of data with neural networks. *Science* **2006**, *313*, 504–507.

9.    Wong, W.K.; Sun, M. Deep Learning Regularized Fisher Mappings. *IEEE Trans. Neural Netw.* **2011**, *22*, 1668–1675.

10.   Han, J.; Zhang, D.; Cheng, G.; Guo, L.; Ren, J. Object Detection in Optical Remote Sensing Images Based on Weakly Supervised Learning and High-Level Feature Learning. *IEEE Trans. Geosci. Remote Sens.* **2015**, *53*, 3325–3337.

11.   Vincent, P.; Larochelle, H.; Lajoie, I.; Bengio, Y.; Manzagol, P.A. Stacked denoising autoencoders: Learning useful representations in a deep network with a local denoising criterion. *J. Mach. Learn. Res.* **2010**, *11*, 3371–3408.

12.   Sun, M.J.; Zhang, D.; Ren, J.C.; Wang, Z.; Jin, J.S. Brushstroke Based Sparse Hybrid Convolutional Neural Networks for Author Classification of Chinese Ink-Wash Paintings. In Proceedings of the 2015 IEEE International Conference on Image Processing (ICIP), Quebec City, QC, Canada, 27–30 September 2015; pp. 626–630.

13.   Krizhevsky, A.; Sutskever, I.; Hinton, G.E. ImageNet classification with deep convolutional neural networks. *Adv. Neural Inf. Proc. Syst.* **2012**, *25*, 1097–1105.

14.   Wang, Y.Y.; Wang, G.H.; Lan, Y.H. PolSAR Image Classification Based on Deep Convolutional Neural Network. *Metall. Min. Ind.* **2015**, *8*, 366–371.

15.   Zhou, Y.; Wang, H.P.; Xu, F.; Jin, Y.Q. Polarimetric SAR Image Classification Using Deep Convolutional Neural Networks. *IEEE Geosci. Remote Sens. Lett.* **2016**, *99*, 1–5.

16.   Aghababaee, H.; Amini, J. Contextual PolSAR image classification using fractal dimension and support vector machines. *Eur. J. Remote Sens.* **2013**, *46*, 317–332.

17.   Margarit, G.; Mallorqui, J.J.; Fabregas, X. Single-Pass Polarimetric SAR Interferometry for Vessel Classification. *Geosci. Remote Sens. IEEE Trans.* **2007**, *45*, 3494–3502.

18.   Bruzzone, L.; Prieto, D.F. Unsupervised retraining of a maximum likelihood classifier for the analysis of multitemporal remote sensing images. *IEEE Trans. Geosci. Remote Sens.* **2001**, *39*, 456–460.

19.   Zhao, Q.; Principe, J.C. Support vector machines for SAR automatic target recognition. *IEEE Trans. Aerosp. Electron. Syst.* **2001**, *37*, 643–654.

20.   Zhang, D.; Chen, S.; Zhou, Z.H. Rapid and brief communication: Learning the kernel parameters in kernel minimum distance classifier. *Pattern Recognit.* **2006**, *39*, 133–135.

21.   Wang, Y.P.; Chen, D.F.; Song, Z.G. Detecting surface oil slick related to gas hydrate/petroleum on the ocean bed of South China Sea by ENVI/ASAR radar data. *J. Asian Earth Sci.* **2013**, *65*, 21–26.

22.   Aplin, P. Image Analysis, Classification and Change Detection in Remote Sensing, with algorithms for ENVI/IDL. *Int. J. Geogr. Inf. Sci.* **2009**, *23*, 129–130.

23.   Lee, J.S.; Grunes, M.R.; Pottier, E. Quantitative comparison of classification capability: Fully polarimetric versus dual and single-polarization SAR. *IEEE Trans. Geosci. Remote Sens.* **2001**, *39*, 2343–2351.

24.   Wang, H.; Zhou, Z.; Turnbull, J.; Song, Q.; Qi, F. Pol-SAR classification based on generalized polar decomposition of mueller matrix. *IEEE Trans. Geosci. Remote Sens.* **2016**, *13*, 565–569.

*applied*
*sciences*

MDPI

Article

# Analysis of Dual- and Full-Circular Polarimetric SAR Modes for Rice Phenology Monitoring: An Experimental Investigation through Ground-Based Measurements

Yuta Izumi [1,*], Sevket Demirci [2], Mohd Zafri bin Baharuddin [3], Tomoro Watanabe [1] and Josaphat Tetuko Sri Sumantyo [1]

[1]   Center for Environmental Remote Sensing, Chiba University, Chiba 2638522, Japan; afaa0135@chiba-u.jp (T.W.); jtetukoss@faculty.chiba-u.jp (J.T.S.S.)
[2]   Electrical and Electronics Faculty of Engineering, Mersin University, Mersin 33343, Turkey; sdemirci@mersin.edu.tr
[3]   Department of Electronics and Communication Engineering, Tenaga national University, Kajang 43000, Malaysia; zafri@uniten.edu.my
*   Correspondence: yutaizumi0927@gmail.com; Tel.: +81-80-1671-9266

Academic Editors: Carlos López-Martínez and Juan Manuel Lopez-Sanchez
Received: 23 February 2017; Accepted: 4 April 2017; Published: 7 April 2017

**Abstract:** Circularly polarized synthetic aperture radar (CP-SAR) is known to be insensitive to polarization mismatch losses caused by the Faraday rotation effect and antenna misalignment. Additionally, the dual-circular polarimetric (DCP) mode has proven to have more polarimetric information than that of the corresponding mode of linear polarization, i.e., the dual-linear polarimetric (DLP) mode. Owing to these benefits, this paper investigates the feasibility of CP-SAR for rice monitoring. A ground-based CP-radar system was exploited, and C-band anechoic chamber data of a self-cultivated Japanese rice paddy were acquired from germination to ripening stages. Temporal variations of polarimetric observables derived from full-circular polarimetric (FCP) and DCP as well as synthetically generated DLP data are analyzed and assessed with regard to their effectiveness in phenology retrieval. Among different observations, the $H/\bar{\alpha}$ plane and triangle plots obtained by three scattering components (surface, double-bounce, and volume scattering) for both the FCP and DCP modes are confirmed to have reasonable capability in discriminating the relevant intervals of rice growth.

**Keywords:** radar polarimetry; synthetic aperture radar (SAR); dual circular polarimetry (DCP); compact polarimetry; rice phenology

---

## 1. Introduction

Conventional polarimetric synthetic aperture radar (SAR) that adopts linearly polarized (LP) antennas on the transmitter and receiver, aptly named as LP-SAR, has already proven its powerful classification ability. Besides, an important performance consideration involves the technical trade-off between the full-linear polarimetric (FLP) and dual-linear polarimetric (DLP) modes. The FLP mode provides complete polarimetric information which enhances the target parameter retrieval and polarimetric discrimination. However, inherent limitations such as reduction of swath width, an increase of system complexity, data rate, and power consumption will be compromised. DLP systems, on the other hand, overcome these limiting factors but they do not afford complete information regarding the full polarization state of the targets. For example, it results in a less accurate alpha angle parameter [1,2] and does not give the possibility to perform three-component (surface, double-bounce, and volume scattering) decomposition. In recent years, to circumvent these drawbacks, the idea to use circular polarization or tilted linear polarization in transmission has emerged, so-called compact polarimetric SAR [3–5]. Three modes have been discussed, i.e., the $\pi/4$ mode [4], the dual-circular polarimetric (DCP) mode [5], and the circular transmit while linear receive (CTLR) mode [3]. These different polarimetric SAR modes are categorized as shown in Figure 1.

More recently, a SAR campaign that utilizes circular polarization in both transmission and reception has been proposed (end-to-end circular polarization) and studied for spaceborne and airborne missions [6]. Although this unique circularly polarized SAR (CP-SAR) system requires almost ideal CP-antennas (i.e., 0 dB of the axial ratio (AR)), which may often be difficult to satisfy, recent advances in antenna technologies can acceptably fulfill this requirement (e.g., AR of 1.1 dB is achieved for RISAT-1) [7]. It is also worth noting that the following advantages of CP-SAR usually outweigh this difficulty in antenna requirement. Circular polarization is already known to be less affected by the Faraday rotation which is a significant problem, especially for lower frequency bands (i.e., L and P) [8]. To be specific, the Faraday rotation effect in the ionosphere does not alter the transmitted polarization which is not the case in linear polarization [9]. Less effect of the interference between direct and reflected signals due to multipath propagation is also expected [6]. In addition, being one of the compact polarimetric modes, the DCP mode of CP-SAR yields more abundant polarimetric information than those of the DLP mode. Furthermore, Guo et al. and Zhang et al. highlighted that the DCP mode is the most suitable configuration among compact polarimetric modes to apply the entoropy-alpha ($H/\bar{\alpha}$) decomposition when this decomposition is performed on a $2 \times 2$ coherency matrix [10,11]. Despite its usefulness, however, CP-SAR has not been practically exploited in Earth observation, and many studies on compact polarimetry reconstruct the compact data by converting from LP-SAR data (i.e., simulated circular polarization data).

Among polarimetric SAR practices, rice monitoring has been one of the important applications since rice is a staple food for almost half the world's population. A less-explored topic in this field is the rice phenology retrieval from the time-series SAR data. This information is of great importance in planning of cultivation practices, yielding estimation and water management. Various analyses of rice monitoring from polarimetric SAR data have been addressed for this purpose [12–16]. Although the classical LP-SAR is the most commonly utilized tool for this task, current studies have been focusing on the compact SAR modes because of their advantages mentioned above [12,14].

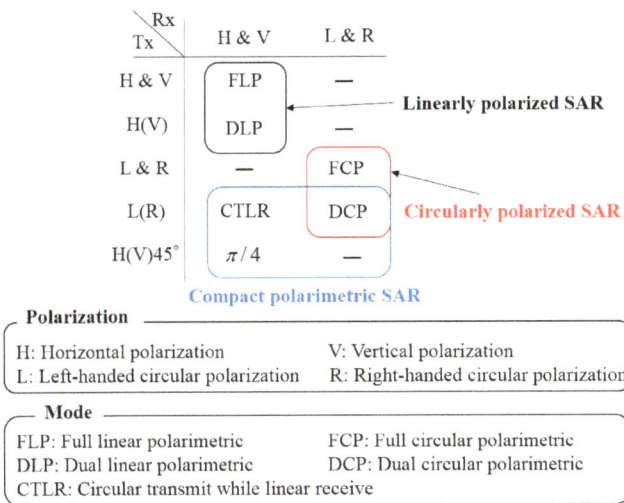

**Figure 1.** Categorization of various polarimetric synthetic aperture radar (SAR) modes.

Due to the growing interest in the use of circular polarization, we investigate, in this paper, the feasibility of the FCP and DCP modes of CP-SAR and its performance on rice phenology monitoring. For this purpose, a ground-based CP-radar system was adopted, and the time-series C-band backscattering data of a Japanese rice paddy (*Oryza sativa* L.) were analyzed. The rice samples were repeatedly observed within an anechoic chamber from germination to ripening stages. To analyze the radar backscatter as a function of growth stages, different polarimetric signatures and target decomposition techniques are exploited for both FCP and DCP data. In addition, $H/\bar{\alpha}$ decomposition for the DLP mode is also examined for comparison purposes.

The paper is organized as follows: the polarimetric decomposition and analysis methodologies for the FCP, DCP, and DLP modes are demonstrated in the next section. Section 3 provides an explanation of the employed ground-based CP-radar system and its polarimetric calibration, the phenological description of the measured rice samples, and the methodology used in data analysis. The experimental results and discussion are given in Section 4. The last section concludes the paper.

## 2. Methodology

Since compact polarimetric SAR systems have been gaining increasing attention, the aim of this study is to assess the performance of the DCP mode by comparing its information content with that of the FCP mode in the case of rice backscattering. For this task, it is required to derive and evaluate the compact versions of eigenvalue/eigenvector-based and model-based target decompositions. The relevant scattering decomposition methods for both the FCP and DCP modes are discussed in the following. The DLP version of eigenvalue/eigenvector-based decomposition is also explained to compare the CP-SAR performance with LP-SAR.

### 2.1. Target Decomposition for FCP Data

Among a substantial number of incoherent decomposition techniques, we chose Cloude–Pottier eigenvector-based $H/\bar{\alpha}$ decomposition (hereafter referred to as $H/\bar{\alpha}$ decomposition) [17] and four-component model-based decomposition for the FCP mode [18,19]. Applying the $H/\bar{\alpha}$ decomposition, the polarimetric entropy $H$ and alpha angle $\alpha$ can be deduced, where $H$ represents scattering randomness, and the $\alpha$ is associated to the corresponding scattering mechanism represented by each eigenvector [17]. The four-component decomposition technique yields four elementary scattering mechanisms, i.e., surface, double-bounce (or dihedral), volume (or multiple), and helix scattering. This methodology was first proposed by Yamaguchi et al. [18], and we adopt herein the improved four-component decomposition proposed by Singh et al. [19] which fully accounts for coherency matrix coefficients.

To apply the aforementioned decomposition theories to FCP data, linear to circular polarization basis transformation is used to obtain first the three-dimensional (3D) Pauli scattering vector in circular basis, given as:

$$\mathbf{k_p} = \frac{1}{\sqrt{2}} \begin{bmatrix} -j2S_{LR} \\ (S_{LL} - S_{RR}) \\ -j(S_{LL} - S_{RR}) \end{bmatrix},$$  (1)

where $S_{xy}$ are the elements of the scattering matrix with $x$ and $y$ denoting the received and transmitted waves respectively, and subscripts L and R represent left-handed circular polarization (LHCP) and right-handed circular polarization (RHCP) respectively. The averaged coherency matrix can then be obtained via the Pauli scattering vector in (1) as:

$$\langle [\mathbf{T_{FCP}}] \rangle = \left\langle \mathbf{k}_p \cdot \mathbf{k}_p^\dagger \right\rangle,$$  (2)

where the superscript $\dagger$ indicates conjugate transpose, and $< \cdot >$ indicates the ensemble averaging operation. This tailored coherency matrix for circular polarization basis makes it possible to adopt decomposition techniques in a similar fashion to linear polarization basis. Under ideal conditions where there are no adverse effects, the information content of FCP and FLP data are identical because each can be transformed to the other via the unitary matrix.

*2.2. Target Decomposition for DCP Data*

Recently, several target decomposition theories for compact polarimetric SAR data have also been proposed. There are three main groups, which differ in the type of the symmetry assumptions that they made about the observed media and/or the matrix used in decomposition: decomposition deduced by the Stokes vector [20]; $2 \times 2$ coherency matrix [10]; and $3 \times 3$ pseudo-coherency matrix [21]. Here, we note that the decomposition deduced from the $3 \times 3$ pseudo matrix was not adopted in this study since the assumed relationship between the correlation coefficient and cross-polarization ratio would not always be valid [21]. Thus, we employed two decomposition approaches, one based on the Stokes vector and the other on the $2 \times 2$ coherency matrix, which carry identical information [20]. The $H/\bar{\alpha}$ decomposition and the three-component decomposition are explained below.

2.2.1. $H/\bar{\alpha}$ Decomposition

We use the unexpanded $2 \times 2$ coherency matrix to implement $H/\bar{\alpha}$ decomposition. With this approach, two eigenvectors corresponding to the first and second dominant eigenvalues can be extracted. The two-dimensional (2D) Pauli vector for circular polarization and the LHCP transmit case is expressed as [10]:

$$\mathbf{k_{DCP}} = \begin{bmatrix} S_{LL} \\ S_{RL} \end{bmatrix}. \tag{3}$$

The coherency matrix is obtained via the Pauli vector in (3) as:

$$\begin{aligned} \langle [\mathbf{T_{DCP}}] \rangle &= \langle \mathbf{k_{DCP}} \cdot \mathbf{k_{DCP}^\dagger} \rangle \\ &= [\ u_1 \quad u_2\ ] \begin{bmatrix} \lambda_1 & 0 \\ 0 & \lambda_2 \end{bmatrix} [\ u_1 \quad u_2\ ]^\dagger. \end{aligned} \tag{4}$$

Here, $\lambda_i$ are the eigenvalues and $u_i$ are the orthogonal eigenvectors of the unitary matrix

$$u_i = e^{j\phi_i}[\ \cos\alpha_i e^{j\delta_i} \quad \sin\alpha_i\ ]^T, \tag{5}$$

where superscript T denotes the transpose operation. The $H$ and mean $\alpha$ ($\bar{\alpha}$) are then given by:

$$H = \sum_{i=1}^{2} P_i(-\log_2 P_i), \tag{6}$$

$$\bar{\alpha} = \sum_{i=1}^{2} P_i \alpha_i, \tag{7}$$

where scattering probabilities are:

$$P_i = \frac{\lambda_i}{\lambda_1 + \lambda_2} (i = 1, 2). \tag{8}$$

The $\bar{\alpha}$ values for the surface, dipole, and double-bounce scattering mechanisms can be calculated from (4), (5), and (7). The corresponding values are shown below together with the FCP values for comparison.

$$\text{surface (DCP)} : 90°, \ \text{dipole (DCP)} : 45°, \ \text{double (DCP)} : 0°,$$

$$\text{surface (FCP)} : 0°, \ \text{dipole (FCP)} : 45°, \ \text{double (FCP)} : 90°.$$

$\alpha$ values of the DCP mode are shown to have symmetric values to those of the FCP mode about the $\alpha = 45°$. Therefore, DCP $\alpha$ values are obtained as $\alpha' = 90° - \alpha$ for comparison with the FCP mode in the results section.

The obtained pairs of $H$ and $\bar{\alpha}$ values are then plotted on the $H/\bar{\alpha}$ 2D plane with some boundaries to clarify the scattering mechanism and feasible region [17]. We perform the analysis on the $H/\bar{\alpha}$ plane for displaying the polarimetric signatures of each rice growth stage. Besides, the FCP and DCP modes have different $H/\bar{\alpha}$ plane plots because the boundaries of each zone differs [11]. Here, our main purpose is to compare the performances of the two modes, thus, we display the $H/\bar{\alpha}$ plane with $\alpha' = 90° - \alpha$ condition for the DCP case. Figure 2 shows the $H/\bar{\alpha}$ planes with different boundaries for each case.

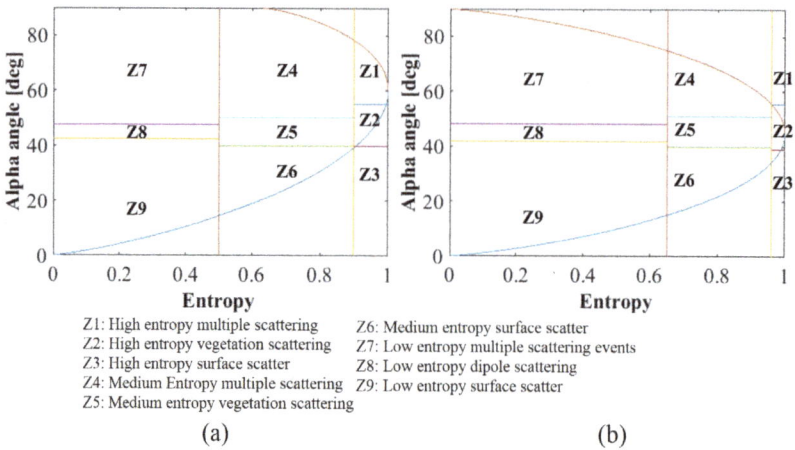

Z1: High entropy multiple scattering
Z2: High entropy vegetation scattering
Z3: High entropy surface scatter
Z4: Medium Entropy multiple scattering
Z5: Medium entropy vegetation scattering
Z6: Medium entropy surface scatter
Z7: Low entropy multiple scattering events
Z8: Low entropy dipole scattering
Z9: Low entropy surface scatter

(a)                                                    (b)

**Figure 2.** The $H/\bar{\alpha}$ plane. (**a**) FCP and FLP; (**b**) DCP.

### 2.2.2. Three-Component Decomposition

Cloude et al. proposed compact three-component decomposition theory using the Stokes vector by considering the relation between the Stokes vector and the $3 \times 3$ coherency matrix [20]. This method extracts three components; namely, surface, double-bounce, and volume scattering power as a function of dominant $\alpha$ and degree of polarization (DoP) which represents the polarized wave ratio of the total receiving power (similar value with entropy).

The Stokes vector for the DCP mode is defined as (LHCP transmit case):

$$\mathbf{g} = \begin{bmatrix} \langle g_0 \rangle \\ \langle g_1 \rangle \\ \langle g_2 \rangle \\ \langle g_3 \rangle \end{bmatrix} = \begin{bmatrix} \langle |S_{RL}|^2 + |S_{LL}|^2 \rangle \\ -2Im\langle S_{RL}S_{LL}^* \rangle \\ 2Re\langle S_{RL}S_{LL}^* \rangle \\ \langle |S_{RL}|^2 - |S_{LL}|^2 \rangle \end{bmatrix}. \tag{9}$$

Three components—the surface ($P_s$), double-bounce ($P_d$), and volume ($P_v$) scattering power—are obtained as:

$$\begin{bmatrix} P_d \\ P_v \\ P_s \end{bmatrix} = \begin{bmatrix} \frac{1}{2}g_0 m(1 - \cos 2\alpha_s) \\ g_0(1 - m) \\ \frac{1}{2}g_0 m(1 + \cos 2\alpha_s) \end{bmatrix}. \tag{10}$$

where $m$ and $\alpha_s$ represent the DoP and dominant $\alpha$ respectively, defined as:

$$m = \frac{1}{g_0}\sqrt{\sum_{i=1}^{3} g_i^2},$$
(11)

$$\alpha_s = \frac{1}{2}\tan^{-1}\left(\frac{\sqrt{g_1^2 + g_2^2}}{\pm g_3}\right) \quad (+ : \text{LHCP transmit} \quad - : \text{RHCP transmit}).$$
(12)

As $\alpha_s$ indicates the dominant $\alpha$, the same $\alpha$ provided by full polarimetric (FP) data can be recovered only when there is a dominant eigenvector in the coherency matrix of FP and a reflection symmetric medium.

### 2.3. Target Decomposition for DLP Data (H/ᾱ Decomposition)

The DLP contains HH/VH or VV/HV scattering matrix coefficients and also yields a $2 \times 2$ coherency matrix—similar to DCP. Consequently, the $H/\bar{\alpha}$ decomposition approach can be directly performed on this coherency matrix [1,22]. Note that, in this study, we use the synthetically generated DLP data obtained by unitary transformation from circular to linear basis:

$$\begin{bmatrix} S_{HH} & S_{HV} \\ S_{VH} & S_{VV} \end{bmatrix} = 2 \begin{bmatrix} 1 & j \\ j & 1 \end{bmatrix}^{-1} \begin{bmatrix} S_{LL} & S_{LR} \\ S_{RL} & S_{RR} \end{bmatrix} \begin{bmatrix} 1 & j \\ j & 1 \end{bmatrix}^{-1}.$$
(13)

The coherency matrix is then expressed by the outer product of the Pauli vector in the same manner as given in (4) (H transmit):

$$\mathbf{k_{DLP}} = \begin{bmatrix} S_{HH} \\ S_{VH} \end{bmatrix},$$
(14)

$$\begin{aligned} \langle[\mathbf{C_{DCP}}]\rangle &= \langle \mathbf{k_{DLP}} \cdot \mathbf{k_{DLP}^\dagger} \rangle \\ &= [\mathbf{U_2}] \begin{bmatrix} \lambda_1 & 0 \\ 0 & \lambda_2 \end{bmatrix} [\mathbf{U_2}]^\dagger, \end{aligned}$$
(15)

where $[\mathbf{U_2}]$ is

$$[\mathbf{U_2}] = \begin{bmatrix} \cos\alpha & -\sin\alpha e^{-i\delta} \\ \sin\alpha e^{i\delta} & \cos\alpha \end{bmatrix}.$$
(16)

To compute $H$ for the DLP mode, the same equation of the DCP mode given in (6) is used. Calculation of $\bar{\alpha}$, however, requires a different process, given as:

$$\bar{\alpha} = P_1\alpha + P_2\left(\frac{\pi}{2} - \alpha\right),$$
(17)

where $P_i$ are calculated in the same manner as (8). Note that the boundary curves of the $H/\bar{\alpha}$ plane for the DLP mode are the same as the DCP curves (see Figure 2b).

## 3. Experimental Scheme

### 3.1. System Description

Circular polarization can be achieved by setting a $90°$ phase difference between H and V polarizations [23]. To introduce this concept, we employed diagonal dual LP-horn antennas with a phase shifter for transmitting and receiving the CP-signal. The vector network analyzer and rotational positioner controlled by the positioner controller were utilized to construct the whole CP-radar system. The experimental geometry illustrated in Figure 3 was adopted, and the rice measurements were conducted within an anechoic chamber to achieve a fully controlled environment. The incidence angle was set to 70°, and the operational frequency was adjusted to 4.5–7.5 GHz (bandwidth is 3 GHz) in C-band. An angular span of 0° to 359° was used to investigate the rice backscattering from all azimuth angles. Therefore, quasi-monostatic antennas always face the target direction.

Before starting experiments, the measurement of the AR parameter of the antennas was needed in order to check the deviation from circular polarization. Theoretically, 0 dB of AR shows ideal circular polarization whereas an infinite value corresponds to ideal linear polarization [23]. An AR value of less than 3 dB is generally considered as an acceptable value for most applications [24], and we follow this definition in this paper. The AR of our CP-antennas was measured inside the anechoic chamber and was found to be under 2.5 dB over the entire operational frequency bandwidth which means that our antennas achieve good circular polarization purity within the investigated bandwidth.

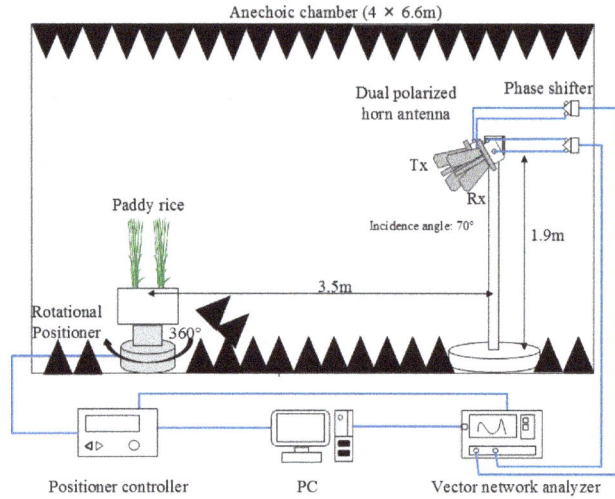

**Figure 3.** Experimental geometry inside an anechoic chamber.

*3.2. Polarimetric Calibration of CP-SAR System*

The scattering matrix of all multi-polarimetric radar systems is easily contaminated by system-introduced distortions such as polarimetric channel imbalances, crosstalk, antenna gain, and impurity of polarization. This contamination inevitably degrades the polarimetric decomposition results, and thus polarimetric calibration should be performed in all cases. Here, we introduce the FLP calibration method proposed by Wiesbeck et al. [25] as a suitable and an effective way for calibrating the ground-based LP-radar systems. To apply this method to CP-radar, an error model in circular polarization basis is constructed [26]:

$$
\begin{bmatrix} S^m_{LL} & S^m_{LR} \\ S^m_{RL} & S^m_{RR} \end{bmatrix} = \begin{bmatrix} R_{LL} & R_{LR} \\ R_{RL} & R_{RR} \end{bmatrix} \begin{bmatrix} S^c_{LL} & S^c_{LR} \\ S^c_{RL} & S^c_{RR} \end{bmatrix} \begin{bmatrix} T_{LL} & T_{LR} \\ T_{RL} & T_{RR} \end{bmatrix} + \begin{bmatrix} I_{LL} & I_{LR} \\ I_{RL} & I_{RR} \end{bmatrix}. \tag{18}
$$

The model includes receive [**R**] and transmit [**T**] distortion matrices; an isolation distortion matrix [**I**]; and correct [**S**$^c$] and measured [**S**$^m$] scattering matrices. With using three types of canonical reflectors, namely, *circular plate, dihedral,* and *45° inclined dihedral,* the error coefficients in distortion matrices [**R**] and [**T**] can be estimated, as explained in detail in our previous study [26]. Empty room calibration is also needed within the steps to extract the isolation distortion matrix [**I**], which can be performed easily by measuring a target free scene.

### 3.3. Phenology Description of Cultivated Rice

Figure 4 shows the photographs and the layout of the rice target used in our experimental validation. A total of eight rice samples were uniformly planted in a rectangular box of dimensions 0.4 × 0.25 × 0.2 m (width × depth × height). The box was made of polystyrene foam for reduction of unwanted echoes and filled with 0.115 m depth of soil and 0.125 m depth of water, as shown in Figure 4b. Thus, the soil was flooded to realize the actual rice field condition. This condition was kept constant throughout the observation period from June to September to collect data that is sensitive only to the rice growth. The non-rice condition was also observed for investigation of the germination stage, where the only flooded soil exists inside the box.

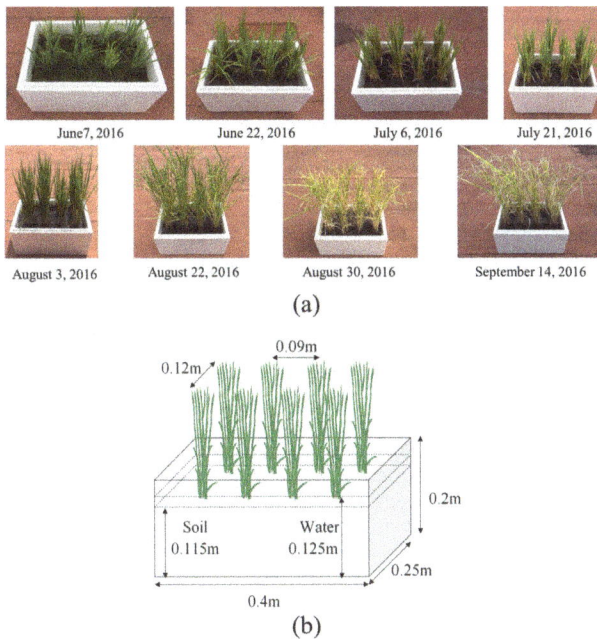

**Figure 4.** Photographs and layout of the rice used in experimental validation. (**a**) Photographs taken on each measurement date within the observation period from 7 June 2016 until 14 September 2016; (**b**) Layout of the eight rice samples uniformly planted within a container box with 0.115 m depth of soil and 0.125 m depth of water.

To describe the different rice phenological stages, we adopted *Biologische Bundesanstalt, Bundessortenamt und CHemische Industrie* (BBCH) decimal scale [27]. BBCH scale provides the description of actual characteristics of an individual plant such as development rate of leaf, tiller, and panicle [13]. Therefore, utilization of this scale is useful to express rice phenology but not for detail morphological expression. BBCH values for all observed rice conditions are given in Table 1 together with day-of-year (DoY) and mean height as a morphological value. Moreover, based on given BBCH codes, five principal stages were designated, namely, Germination (stage 1), Tillering (stage 2), Stem elongation (stage 3), Booting (stage 4), and Ripening (stage 5) stages, where we referred to [27].

**Table 1.** BBCH code and phenology stage of each observed data set.

| Date | DoY (Day of Year) | Mean Height (cm) | BBCH Code | Phenological Stage |
|---|---|---|---|---|
| Soil and water | NA | NA | 0 | 1: Germination |
| 7 June 2016 | 159 | 19 | 21-29 | 2: Tillering |
| 22 June 2016 | 174 | 27 | 21-29 | 2: Tillering |
| 6 July 2016 | 188 | 34 | 30-39 | 3: Stem elongation |
| 21 July 2016 | 203 | 42 | 30-39 | 3: Stem elongation |
| 3 August 2016 | 216 | 49 | 41-49 | 4: Booting |
| 22 August 2016 | 235 | 52 | 83-85 | 5: Ripening |
| 30 August 2016 | 243 | 52 | 87-89 | 5: Ripening |
| 14 September 2016 | 258 | 45 | 93 | 5: Ripening |

Our rice plants were discovered to have an increasing number of tillers until stage 2. Finally, an average of 17 tillers in each stock was observed by 6 July. The increase in the number of tillers was stopped at the end of stage 2 and samples began to initiate panicles inside the stem during stage 3. Just before heading at stage 4, panicles went up and started to come out of the stems. Finally, we found head emergence on 6 August for this type of rice sample.

*3.4. Methodology of Data Analysis*

As explained in Section 3.1, we adopted a 2D inverse SAR (ISAR) data collection geometry with a fixed incidence angle. Thus, the frequency domain backscatter data of rice samples were acquired for the complete azimuth angles from $0°$ to $359°$. Figure 5 shows the reconstructed images of one sample of data observed on 30 August 2016, for LL, RL, and RR polarizations. A spherical back-projection algorithm is used to process this wide-angle data [28]. From the reflectivity images in Figure 5, it can be clarified that all rice stocks are clearly identified for each polarization thanks to the high-resolution capability of our system. In practical rice monitoring applications, however, the resulted SAR images cannot usually maintain such a high-resolution feature, and a single resolution cell consists of the superposition of different scattering contributions from a few rice plants. Thus, for our situation, all eight rice plants should be confined inside a single resolution cell for a reasonable analysis of rice scattering mechanisms, as Sagues et al. mentioned [29]. For this reason, our analyses are not performed on image data but the scatterometric data obtained by ensemble averaging of the frequency domain data along with all the azimuth angles (0–359°) and frequencies (4.5–7.5 GHz). As a result, the whole scattering behavior is combined into a single coherency matrix as also adopted in [29–31], and the decomposition theories explained in Section 2 are applied to this matrix.

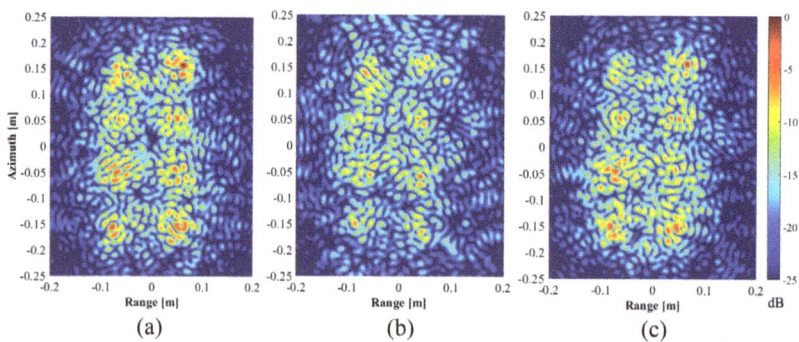

**Figure 5.** Reconstructed circularly polarized SAR (CP-SAR) images for the rice observed on 30 August 2016. The images are normalized to the maximum value of three images. (**a**) LL polarization; (**b**) RL polarization; (**c**) RR polarization.

## 4. Rice Monitoring Results and Discussion

In the following subsection, the backscattering coefficients and the polarimetric decomposition results deduced by the averaged single coherency matrix will be presented for the extraction of physical scattering mechanisms of rice growth.

### 4.1. Backscattering Coefficient

Three scattering matrix data; $S_{LL}$, $S_{RL}$, and $S_{RR}$ were collected by assuming $S_{LR} = S_{RL}$ since the reciprocity theorem almost holds true for our quasi-monostatic setup. Figure 6 shows the intensity of backscattering coefficients (LL, RL, and RR) as a function of DoY. The backscattering coefficients are normalized to a maximum value of the whole observation period. From Figure 6, we can see that the backscattering coefficients for all three polarizations increase as rice plants grow until DoY 235 because these are related to the leaf area index, rice freshness, and rice height [32], where our rice plants stop to increase those heights from DoY 235, shown in Table 1. Before rice emergence from the soil (non-rice observation), the backscattering intensity exhibits very low values as a result of the specular reflection from the flooded ground.

The sense of CP-waves is reversed when they are reflected from flat surfaces and smoothly curved spherical reflectors. Therefore, cross- and co-polarization indicate odd- and even-bounce scattering mechanisms respectively. Based on this fact, it is seen from Figure 6 that the surface (cross-polarization) scattering produces relatively higher intensities than the double-bounce (co-polarization) scattering until the end of stage 3 and the opposite is true for later stages. Nevertheless, the maximum difference between polarimetric channels is around 2 dB which is not so significant.

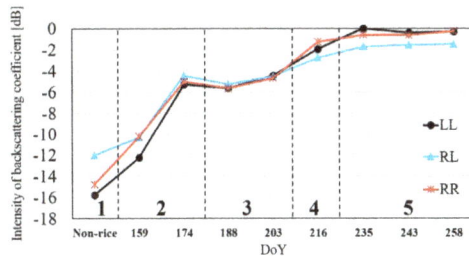

**Figure 6.** Backscattering coefficients of LL, RL, and RR polarization.

### 4.2. H/ᾱ Decomposition

The entropy and ᾱ values for the FCP, DCP, and DLP modes are plotted in Figure 7. Note that the ᾱ for the DCP mode was treated with the condition $\bar{\alpha}' = 90° - \bar{\alpha}$ for comparison purposes as discussed in Section 2.2.1.

The entropy values shown in Figure 7a exhibit relatively lower values at stage 1, because of the non-planted soil condition which produces highly deterministic scattering. Right after vegetation starts to grow, other scattering mechanisms such as double-bounce and volume scattering make entropy higher. Except for stage 1, entropy values fluctuate within the ranges 0.95–1 for FCP, 0.8–0.9 for DCP, and 0.4–0.6 for the DLP mode. It can also be noted that the entropy values for FCP and DCP represent a similar progressive trend and around 0.12 discrepancy between them. On the other hand, entropy for DLP ranges between 0.4 and 0.6 which is comparably lower than those of FCP and DCP. Furthermore, an abrupt decrease at DoY 203 is also observed in the DLP mode, which is not the case in the FCP and DCP modes.

The ᾱ characteristic is shown in Figure 7b. FCP and DCP patterns are approximately identical except for non-rice and DoY 159 observations which differ by about ~5°. Despite this, each pattern shows an increasing trend, indicating a progression from surface scattering to dipole-like scattering.

On the contrary, $\bar{\alpha}$ of DLP reveals a notably different result. Stage 1 measured a high $\bar{\alpha}$ of ~78°, which is regarded as multiple scattering. The value then stabilizes to around ~75° on average.

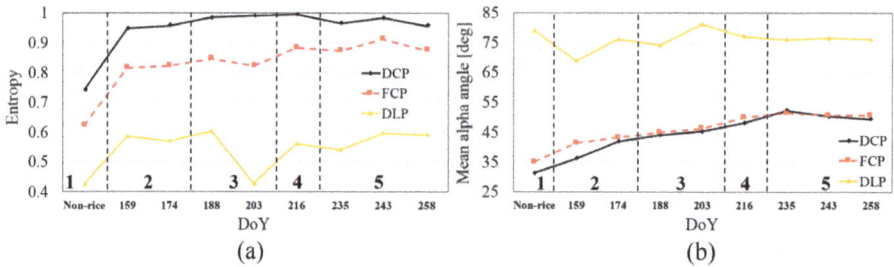

**Figure 7.** $H/\bar{\alpha}$ decomposition results of the FCP, DCP, and DLP modes. (**a**) Entropy; (**b**) Mean alpha.

To gain a deeper understanding of the scattering mechanisms contained in FCP and DCP data, the independent components of the $H/\bar{\alpha}$ decomposition are investigated separately. Firstly, the appearance probabilities $P_i$ for each scattering type defined by the associated eigenvector are calculated (see (6) and (8) for the DCP case) to interpret the entropy plots in Figure 7a. Note that the decomposition of FCP and DCP data produces three and two eigenvectors/probabilities, respectively. Also, the probability values are constrained by the expressions, $P_1 + P_2 + P_3 = 1$ and $P_1 \geq P_2 \geq P_3$ for FCP, and similarly for DCP. The variation of these probabilities as a function of DoY are displayed in Figure 8. The probability $P_1$ for DCP data shows a similar trend to that of FCP over the whole observation period but experiences slightly higher value (~10%) for stages 3 and 4. From Figure 8a, an almost constant and relatively large $P_3$ (recessive scattering) (~10%) is observed for FCP data. This is the case when more than three scattering mechanism components contribute to the receiving signals. Usually, complex targets such as forested areas give rise to this type of multi-scattering. If the third eigenvalue corresponding to $P_3$ strongly affects the receiving signal, the difference between the entropy values between FCP and DCP becomes higher because entropy is formulated by summation of each independent scattering contribution which corresponds to scattering probability. As Cloude et al. pointed out in [20], compact mode systems typically produce higher entropy compared to the FP system.

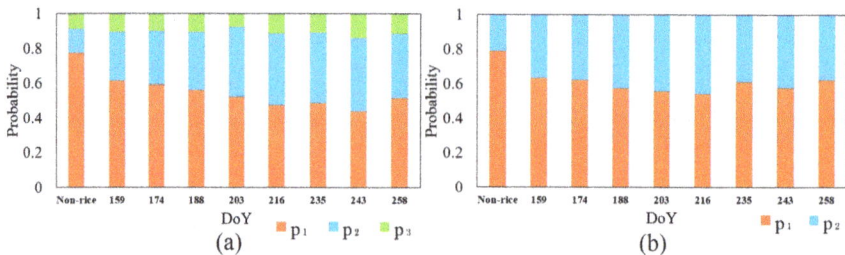

**Figure 8.** Scattering probabilities for the FCP and DCP modes. (**a**) FCP mode; (**b**) DCP mode.

Secondly, $\bar{\alpha}$ results given in Figure 7b are investigated in terms of their components. For each eigenvector $u_i$, the corresponding $\alpha_i$ is extracted by using $\alpha_i = \arccos(|u_{1i}|)$ where $|u_{1i}|$ is the absolute value of the first element of the eigenvector. From the results given in Figure 9, we notice that $\alpha_1$ and $\alpha_2$ for DCP indicate different trends and values from FCP values. It is observed that $\alpha_1$ for FCP is close to 45° at stage 4 and fluctuates within the range of 40–65° at stage 5, whereas $\alpha_1$ for DCP is close to 45° at stage 3 and fluctuates within the range of 60–85° at stages 4 and 5. Thus, these results confirm the necessity of statistical interpretation in terms of $\bar{\alpha}$ values and/or $H/\bar{\alpha}$ space for the discrimination of physical scattering mechanisms.

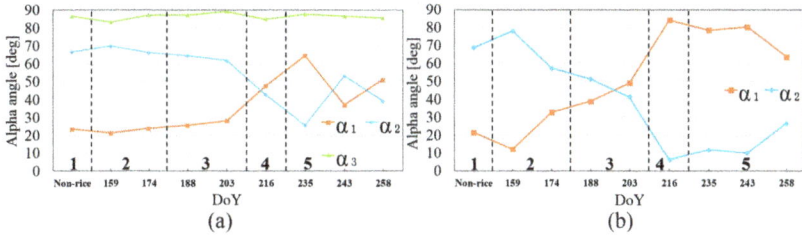

**Figure 9.** Alpha angle corresponding to each eigenvector. (**a**) FCP mode; (**b**) DCP mode.

Figure 10 displays the $H/\bar{\alpha}$ 2D plane plots obtained from the entropy and $\bar{\alpha}$ values for FCP (Figure 10a), DCP (Figure 10b), and DLP (Figure 10c) measurements. In the FCP mode, the distributions mainly lie in zones 6, 5, and 4, indicating a vegetation-type scattering event. Roughly, four different clusters can be identified: germination stage, tillering stage, stem elongation stage, and booting plus ripening stage, and thus the booting and ripening stages cannot be separated effectively from FCP data. The results for the DCP mode in Figure 10b also reveal a broadened $H/\bar{\alpha}$ pattern, but the two observations (DoY 159 and 174) in the tillering stage are moved away from each other more than the FCP case. Visually, the DCP mode affords better separation between booting and ripening stages compared to the FCP mode, but this cannot be regarded as better capability because the DCP mode has less information content than the FCP mode. Although the $H/\bar{\alpha}$ plane plots for the FCP and DCP modes exhibit some different features, each mode yields satisfactory discrimination of the phenological intervals. In contrast, the $H/\bar{\alpha}$ plane plots for the DLP mode cannot achieve adequate discrimination capability for our rice targets, since almost all intervals are not resolved successfully, as seen from Figure 10c.

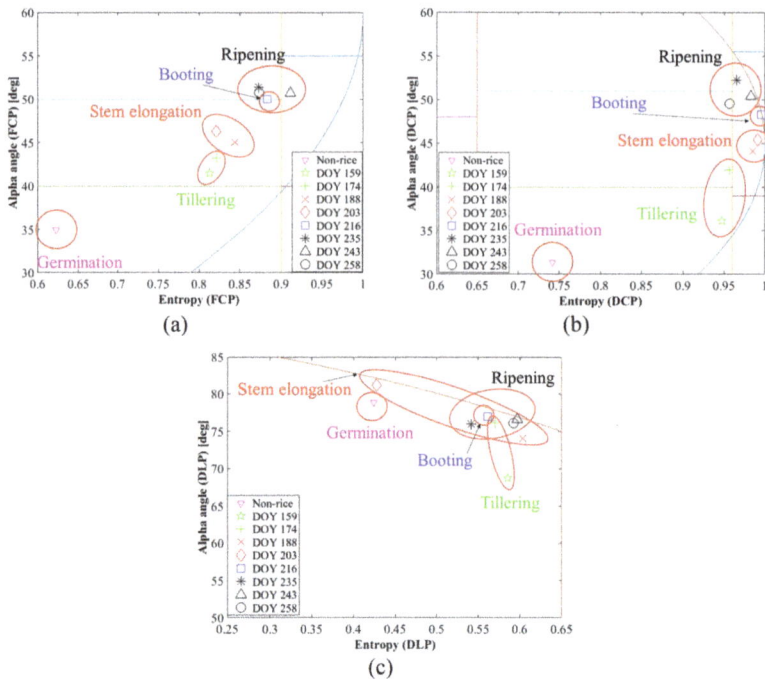

**Figure 10.** The $H/\bar{\alpha}$ 2D plane. (**a**) FCP mode; (**b**) DCP mode; (**c**) DLP mode.

It is also worth noting that entropy and $\bar{\alpha}$ values for crop backscattering usually increase with increasing incidence angle. For lower incidence angles, surface reflection from soil is dominant whereas, for high incidence angles (e.g., greater than 20°), backscattering from vegetation stems and leaves make dipole-like scattering dominant, as demonstrated in [30]. Therefore, because of our high incidence angle (70°) feature, relatively higher $H$ and $\bar{\alpha}$ values from our experiment than the usual spaceborne SAR incidence angle (common operational mode: <70°) can be expected.

### 4.3. Four- and Three-Component Decomposition

The four- and three-component decomposition results for the FCP and DCP modes are shown in Figure 11 respectively, where $P_s$, $P_d$, $P_v$, and $P_c$ show surface, double-bounce, volume, and helix scattering respectively. The results are normalized to the maximum value of both modes. In Figure 11a, the double-bounce and surface scattering contributions of four-component decomposition demonstrate a similar evolutionary trend as the backscattering coefficients of co- and cross-polarization in Figure 6. This similarity proves that co- and cross-polarization indicate even and odd bounce scattering mechanisms respectively, as mentioned in Section 4.1. Figure 11a also reveals a stronger volume scattering component than other scattering mechanisms at stages 4 and 5 as well as its evolutionary trend during rice growth. Relatively low helix scattering is also shown, as expected, from vegetation.

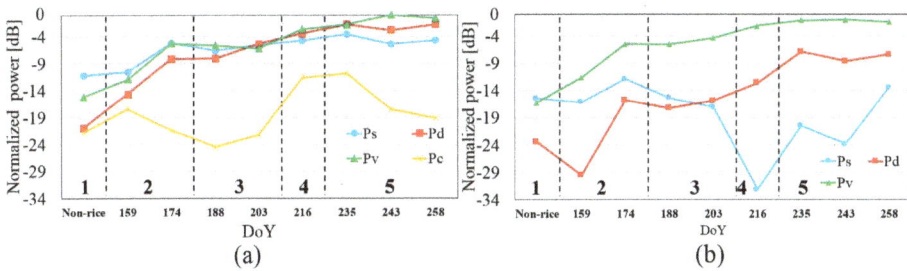

**Figure 11.** Four-/three-component decomposition results. (**a**) Four-component decomposition results for the FCP mode; (**b**) Three-component decomposition results for the DCP mode.

When we compare the results of both modes, we see that both volume scattering components result in a similar evolutionary trend and value. In contrast, the surface and double-bounce scattering of the DCP mode show lower power than the FCP mode. This error might be attributed to dominant $\alpha$ values because surface and double-bounce scattering are a function of dominant $\alpha$ values, shown in (12). Since we found the dominant $\alpha$ difference between the FCP and DCP modes in Figure 9, this error obviously affects the power level.

To investigate the relative contribution of the scattering component, we present the rate of each scattering mechanism on a triangle plot, which is a similar analysis to [33], shown in Figure 12a,b for the FCP and DCP modes respectively. Note that we exclude the helix scattering contribution from the FCP mode for comparison between both modes. Figure 12a shows similar results to $H/\bar{\alpha}$ display, where the triangle plot for the FCP mode loosely falls into four groups: germination, tillering, stem elongation, and a mix of booting and ripening stages. Therefore, booting and ripening stages still cannot be clearly separated. Moreover, we notice that double-bounce scattering gradually increases as rice grows (from 7% to 36%), while vice versa, the surface scattering decreases in the FCP mode. This situation can also be seen in the DCP mode, where double-bounce scattering increases from 7.8% to 20%, as depicted in Figure 12b. However, as indicated above, the DCP mode shows a relatively stronger volume scattering component (lower surface and double-bounce scattering contribution) in Figure 12b. Therefore, three-component decomposition for the compact mode should be improved to better approximate the FP mode results.

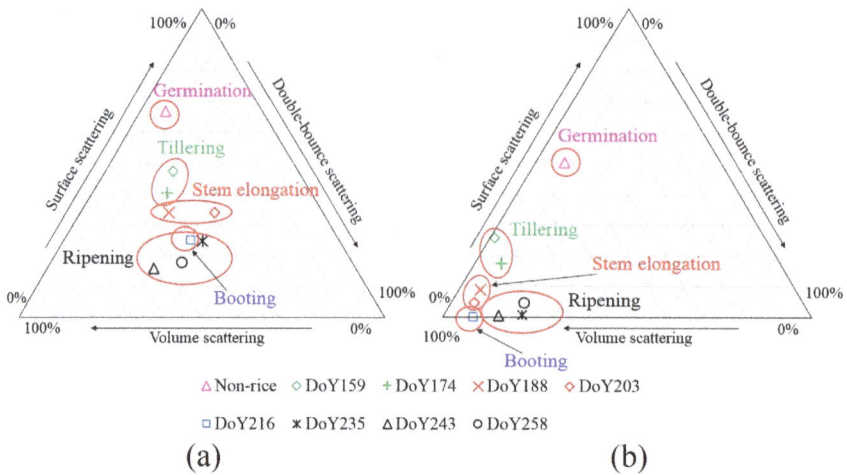

**Figure 12.** Relative contribution of surface, double-bounce, and volume scattering components on the triangle plot. (**a**) FCP mode; (**b**) DCP mode.

## 5. Conclusions

This work investigated the FCP and DCP modes of CP-SAR (end-to-end circular polarization system) through rice growth monitoring. For this purpose, self-cultivated rice plants from germination to ripening stages were analyzed by means of frequency domain data measurement using a ground-based CP-radar system. We applied the $H/\bar{\alpha}$ and four-/three-component polarimetric decomposition techniques to measurement data.

The results for the two types of polarimetric decomposition can be summarized as:

- The $H/\bar{\alpha}$ 2D plane showed a satisfactory clear pattern of each stage of rice growth and yielded rough discriminating capability. However, booting and ripening stages could not be separated.
- The DCP mode exhibited better classification capability than DLP mode on the $H/\bar{\alpha}$ 2D plane.
- The four-/three-component decomposition results demonstrated a similar trend of surface and double-bounce scattering as backscattering coefficients of cross- and co-polarization respectively.
- The triangle plot of relative scattering components contribution showed adequate classification capability, similar to the $H/\bar{\alpha}$ plane.

The comparison results between FCP and DCP modes can be stated as:

- Entropy showed a difference of ∼0.12 between each other, but overall evolution exhibited similarity. $\bar{\alpha}$ values almost coincided with each other except for non-rice and DoY 159 observations, but dominant and second dominant $\alpha$ resulted in differences.
- The volume scattering component yielded a very similar trend and value. Surface and double-bounce scattering presented different power to each other.

Overall, results for both the FCP and DCP modes of CP-SAR demonstrated adequate rice phenology classification capability. Differences in some polarimetric decomposition parameters between the FCP and DCP modes were noted. Moreover, the DCP mode yielded similar performance to the FCP mode about the $H/\bar{\alpha}$ decomposition and volume scattering, and better classification performance of rice phenology compared to the DLP mode.

This work showed the first results of the ground-based CP radar system for long-term rice observation. We anticipate that this study will contribute to the future development and implementation of CP-SAR systems for space-borne and airborne missions, to be used for precise and efficient monitoring of vegetation and ground surface by combining the advantages of circular polarization and the compact mode.

**Acknowledgments:** This work was supported in part by the European Space Agency Earth Observation Category 1 under Grant 6613, by the 4th Japan Aerospace Exploration Agency (JAXA) ALOS Research Announcement under Grant 1024, by the 6th JAXA ALOS Research Announcement under Grant 3170, by the Japanese Government National Budget (Special Budget for Project) FY 2015 under Grant 2101, Taiwan National Space Organization under Grant NSPO-S-105096; Indonesian Bhimasena, and by the Chiba University Strategic Priority Research Promotion Program FY 2016.

**Author Contributions:** Yuta Izumi led this work. He proposed the idea, processed data, compiled results and wrote this manuscript. Sevket Demirci and Josaphat Tetuko Sri Sumantyo supervised with some fruitful ideas to make this work meaningful. Mohd Zafri bin Baharuddin and Tomoro Watanabe arranged the SAR-system and experiment together with Yuta Izumi.

**Conflicts of Interest:** The authors declare no conflict of interest.

## References

1. Dhar, A.T.; Gray, B.D.; Menges, C.C. Comparison of dual and full polarimetric entropy/alpha decompositions with TerraSAR-X, suitability for use in classification. In Proceedings of the Geoscience and Remote Sensing Symposium (IGARSS), Vancouver, BC, Canada, 24–29 July 2011; pp. 456–458.

2. Sugimoto, M.; Ouchi, K.; Nakamura, Y. On the similarity between dual-and quad-eigenvalue analysis in SAR polarimetry. *Remote Sens. Lett.* **2013**, *4*, 956–964.

3. Raney, R.K. Hybrid-polarity SAR architecture. *IEEE Trans. Geosci. Remote Sens.* **2007**, *45*, 3397–3404.

4. Souyris, J.C.; Imbo, P.; Fjortoft, R.; Mingot, S.; Lee, J.S. Compact polarimetry based on symmetry properties of geophysical media: The $\pi/4$ mode. *IEEE Trans. Geosci. Remote Sens.* **2005**, *43*, 634–646.

5. Stacy, N.; Preiss, M. Compact polarimetric analysis of X-band SAR data. In Proceedings of the European Conference on Synthetic Aperture Radar (EUSAR), Dresden, Germany, 16–18 May 2006.

6. Tetuko, S.S.J.; Koo, V.C.; Lim, T.S.; Kawai, T.; Ebinuma, T.; Izumi, Y.; Baharuddin, M.Z.; Gao, S.; Ito, K. Development of circularly polarized synthetic aperture radar on-board UAV JX-1. *Int. J. Remote Sens.* **2017**, *38*, 1–12.

7. Rao, Y. S.; Meadows, P.; Kumar, V. Evaluation of RISAT-1 compact polarization data for calibration. In Proceedings of the Geoscience and Remote Sensing Symposium (IGARSS), Beijing, China, 10–15 July 2016; pp. 3250–3253.

8. Freeman, A. Calibration of linearly polarized polarimetric SAR data subject to Faraday rotation. *IEEE Trans. Geosci. Remote Sens.* **2004**, *42*, 1617–1624.

9. Souyris, J.C.; Stacy, N.; Ainsworth, T.; Lee, J.S.; Dubois-Fernandez, P. SAR compact polarimetry (CP) for earth observation and planetology: Concept and challenges. In Proceedings of the PolInSAR, Frascati, Italy, 22–26 January 2007.

10. Guo, R.; Liu, Y.B.; Wu, Y.H.; Zhang, S.X.; Xing, M.D.; He, W. Applying $H–\alpha$ decomposition to compact polarimetric SAR. *IET Radar Sonar Navig.* **2012**, *6*, 61–70.

11. Zhang, H.; Xie, L.; Wang, C.; Wu, F.; Zhang, B. Investigation of the Capability of $H–\alpha$ Decomposition of Compact Polarimetric SAR. *IEEE Geosci. Remote Sens. Lett.* **2014**, *11*, 868–872.

12. Lopez-Sanchez, J.M.; Vicente-Guijalba, F.; Ballester-Berman, J.D.; Cloude, S.R. Polarimetric response of rice fields at C-band: Analysis and phenology retrieval. *IEEE Trans. Geosci. Remote Sens.* **2014**, *52*, 2977–2993.

13. Lopez-Sanchez, J.M.; Cloude, S.R.; Ballester-Berman, J.D. Rice phenology monitoring by means of SAR polarimetry at X-band. *IEEE Trans. Geosci. Remote Sens.* **2012**, *50*, 2695–2709.

14. Yang, Z.; Li, K.; Liu, L.; Shao, Y.; Brisco, B.; Li, W. Rice growth monitoring using simulated compact polarimetric C band SAR. *Radio Sci.* **2014**, *49*, 1300–1315.

15. Hayashi, N.; Sato, M. Measurement and analysis of paddy field by polarimetric GB-SAR. In Proceedings of the Geoscience and Remote Sensing Symposium (IGARSS), Cape Town, South Africa, 12–17 July 2009; pp. IV358–IV361.

16. Li, K.; Brisco, B.; Yun, S.; Touzi, R. Polarimetric decomposition with RADARSAT-2 for rice mapping and monitoring. *Can. J. Remote Sens.* **2012**, *38*, 169–179.

17. Cloude, S.R.; Pottier, E. An entropy based classification scheme for land applications of polarimetric SAR. *IEEE Trans. Geosci. Remote Sens.* **1997**, *35*, 68–78.

18. Yamaguchi, Y.; Moriyama, T.; Ishido, M.; Yamada, H. Four-component scattering model for polarimetric SAR image decomposition. *IEEE Trans. Geos. Remote Sens.* **2005**, *43*, 1699–1706.
19. Singh, G.; Yamaguchi, Y.; Park, S.E. General four-component scattering power decomposition with unitary transformation of coherency matrix. *IEEE Trans. Geosci. Remote Sens.* **2013**, *51*, 3014–3022.
20. Cloude, S.R.; Goodenough, D.G.; Chen, H. Compact decomposition theory. *IEEE Geosci. Remote Sens. Lett.* **2012**, *9*, 28–32.
21. Nord, M.E.; Ainsworth, T.L.; Lee, J.S.; Stacy, N.J. Comparison of compact polarimetric synthetic aperture radar modes. *IEEE Trans. Geosci. Remote Sens.* **2009**, *47*, 174–188.
22. Cloude, S. The dual polarization entropy/alpha decomposition: A PALSAR case study. In Proceedings of the PolInSAR, Frascati, Italy, 22–26 January 2007.
23. Stutzman, W.L. *Polarization in Electromagnetic Systems*; Artech House: Boston, MA, USA; London, UK, 1993.
24. Gao, S.; Luo, Q.; Zhu, F. *Circularly Polarized Antennas*; Wiley: Hoboken, NJ, USA, 2013.
25. Wiesbeck, W.; Kahny, D. Single reference, three target calibration and error correction for monostatic, polarimetric free space measurements. *Proc. IEEE* **1991**, *79*, 1551–1558.
26. Izumi, Y.; Demirci, S.; Baharuddin, M.Z.; Waqar, M.M.; Sumantyo, J.T.S. The development and comparison of two polarimetric calibration techniques for ground-based circularly polarized radar system. *Prog. Electromagn. Res. B* **2017**, *73*, 79–93.
27. Lancashire, P.D.; Bleiholder, H.; Boom, T.V.D.; Langeluddeke, P.; Stauss, R.; Weber, E.; Witzenberger, A. A uniform decimal code for growth stages of crops and weeds. *Ann. Appl. Biol.* **1991**, *119*, 561–601.
28. Demirci, S.; Yigit, E.; Ozdemir, C. Wide-field circular SAR imaging: An empirical assessment of layover effects. *Microw. Opt. Technol. Lett.* **2015**, *57*, 489–497.
29. Sagues, L.; Lopez-Sanchez, J.M.; Fortuny, J.; Fabregas, X.; Broquetas, A.; Sieber, A.J. Indoor experiments on polarimetric SAR interferometry. *IEEE Trans. Geosci. Remote Sens.* **2000**, *38*, 671–684.
30. Lopez-Sanchez, J.M.; Fortuny-Guasch, J.; Cloude, S.R.; Sieber, A.J. Indoor polarimetric radar measurements on vegetation samples at L, S, C and X band. *J. Electromagn. Waves Appl.* **2000**, *14*, 205–231.
31. Zhou, Z.S.; Cloude, S. Structural parameter estimation of australian flora with a ground-based polarimetric radar interferometer. In Proceedings of the Geoscience and Remote Sensing Symposium (IGARSS), Denver, CO, USA, 31 July–4 August 2006; pp. 71–74.
32. Inoue, Y.; Sakaiya, E.; Wang, C. Capability of C-band backscattering coefficients from high-resolution satellite SAR sensors to assess biophysical variables in paddy rice. *Remote Sens. Environ.* **2014**, *140*, 257–266.
33. Yonezawa, C.; Negishi, M.; Azuma, K.; Watanabe, M.; Ishitsuka, N.; Ogawa, S.; Saito, G. Growth monitoring and classification of rice fields using multitemporal RADARSAT-2 full-polarimetric data. *Int. J. Remote Sens.* **2012**, *33*, 5696–5711.

![applied sciences logo] *applied sciences*

MDPI

*Article*

# A Multi-Year Study on Rice Morphological Parameter Estimation with X-Band Polsar Data

**Onur Yuzugullu [1,\*], Esra Erten [2] and Irena Hajnsek [1,3]**

[1]   Institute of Environmental Engineering ETH Zurich, 8093 Zurich, Switzerland; irena.hajnsek@dlr.de
[2]   Faculty of Civil Engineering, Istanbul Technical University, 34469 Istanbul, Turkey; eerten@itu.edu.tr
[3]   Microwaves and Radar Institute, German Aerospace Center (DLR), 82234 Oberpfaffenhofen, Germany
[\*]   Correspondence: onuryuzugullu@live.com

Academic Editors: Carlos López-Martínez and Juan Manuel Lopez-Sanchez
Received: 29 March 2017; Accepted: 8 June 2017; Published: 9 June 2017

**Abstract:** Rice fields have been monitored with spaceborne Synthetic Aperture Radar (SAR) systems for decades. SAR is an essential source of data and allows for the estimation of plant properties such as canopy height, leaf area index, phenological phase, and yield. However, the information on detailed plant morphology in meter-scale resolution is necessary for the development of better management practices. This letter presents the results of the procedure that estimates the stalk height, leaf length and leaf width of rice fields from a copolar X-band TerraSAR-X time series data based on a priori phenological phase. The methodology includes a computationally efficient stochastic inversion algorithm of a metamodel that mimics a radiative transfer theory-driven electromagnetic scattering (EM) model. The EM model and its metamodel are employed to simulate the backscattering intensities from flooded rice fields based on their simplified physical structures. The results of the inversion procedure are found to be accurate for cultivation seasons from 2013 to 2015 with root mean square errors less than 13.5 cm for stalk height, 7 cm for leaf length, and 4 mm for leaf width parameters. The results of this research provided new perspectives on the use of EM models and computationally efficient metamodels for agriculture management practices.

**Keywords:** polarimetry; SAR; precision agriculture; rice monitoring; stochastic optimization; metamodels; radiative transfer models; electromagnetic scattering models

## 1. Introduction

Rice is the main source of food and income for several highly populated countries. The increasing population and limited arable lands bring out the need for higher yields that depends on the development of better management practices. The traditional method monitoring by visual inspection is not possible for kilometer-square areas. For such large scales, remote sensing based methods are good alternatives. Among different data sources, information provided by Synthetic Aperture Radar (SAR) is advantageous with its sensitivity to geometric and dielectric properties of the objects and its availability in all light and weather conditions. Therefore, SAR is a valuable tool for monitoring the rice fields [1].

Understanding the phenological evolution of rice fields is important to develop effective management strategies. In the literature, several SAR data based algorithms have been developed to investigate the phenological evolution of rice fields including their canopy height, growth stage, leaf area index and yields. In SAR based rice monitoring, one way of monitoring the phenological cycle is the use of different SAR data analysis techniques including temporal trend analysis [2,3], interferometric analysis [4,5] and polarimetric interferometry analysis [6,7]. The other approach employs EM models to simulate the polarimetric parameters from plant properties [5,8,9].

*Appl. Sci.* **2017**, *7*, 602

This letter presents the inversion results of the metamodel-driven EM model for the estimations of stalk height, leaf length, and leaf width parameters from copolar X-band TerraSAR-X time series data, by following the methodology given in Figure 1, detailed and presented in [10] for a single year and growth stage parameter, specifically BBCH. Unlike [10], this study focuses on the feasibility of the proposed approach for morphology parameter estimation under different conditions. The chosen EM model [11] uses the biophysical properties of rice plants to simulate the backscattering intensities ($\bar{\sigma}^o$) with Monte-Carlo simulations. The computation costs related to the multi-dimensional algorithm and the Monte-Carlo simulations were reduced by introducing metamodels, which mimics the EM model after trained for once. The stability of the inversion was tested over a dataset of three different cultivation periods (2013–2015), which includes data from rice fields under different agricultural and environmental conditions.

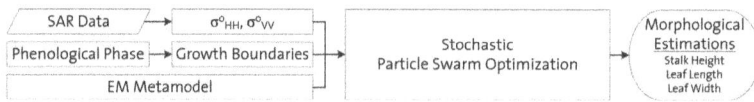

**Figure 1.** Block-diagram of the proposed stochastic inversion of metamodel-driven electromagnetic scattering (EM) model.

This paper has four sections. It starts with the proposed methodology in Section 2 by providing the EM model, its metamodel, and the inversion procedure. Sections 3 and 4 present the SAR and ground data followed by the inversion analysis results. The overall summary is provided in Section 5.

## 2. Methodology

### 2.1. Theoretical EM Model and Its Metamodel

In this study, the EM model [11], $\mathcal{M}(\boldsymbol{\xi})$, is employed to simulate the backscattering intensities, $\bar{\sigma}^o$, from a set of rice plant morphology parameters, $\boldsymbol{\xi}$, through Monte Carlo simulations for varying scatterer locations. The $\boldsymbol{\xi}$ set includes stalk dimensions (height and diameter), leaf dimensions (length and width), panicle dimensions (length and width), and their structural densities. The simulations are done for a unit area $A$, having randomly placed non-overlapping cylindrical stalks with a specific height and diameter. Each stalk is modeled to have elliptical leaves with a fixed length and width. In this study, we assumed flooded ground and plant components with fixed complex dielectric constants for the complete growth cycle by relying on the sensitivity analysis of the EM model [12]. The $\mathcal{M}(\boldsymbol{\xi})$, given in (1), provides the relation between an incident $\bar{E}^i$ and a scattered wave $\bar{E}^s$ through the coherent sum of four different scattering mechanisms ($S_n$), as shown in Figure 2.

$$\bar{\sigma}^o_{qq} = \mathcal{M}(\boldsymbol{\xi}) = \frac{4\pi r^2}{A} \frac{\langle |E^s_q|^2 \rangle}{|E^i_q|^2} = \left\langle \left| \frac{e^{ikr}}{r} \left( \sum_{n=1}^{4} S_n \right) \right|^2 \right\rangle \tag{1}$$

**Figure 2.** The four different scattering mechanisms considered within the EM model [11]. $S_1$: Direct scattering from the scatterers, $S_2$: Scattering from the canopy followed by reflection from the ground, $S_3$: Reflection from the ground followed by scattering from the canopy, $S_4$: Reflection from the ground followed by scattering from the canopy and followed by reflection from the ground.

In the EM model given in (1), $q$ and $p$ subscripts correspond to transmitted and received horizontal (H) and vertical (V) linear polarization channels. The parameters $k$ and $r$ represent the free-space wavenumber and the distance between the sensor and the target, respectively. The $\bar{\sigma}^0_{qq}$ for the different polarimetric channels, $qq$, are approximated from the ratio between $E^s_q$ and $E^i_q$.

The $\mathcal{M}(\xi)$ has high computation cost due to its multi-dimensional algorithm and Monte Carlo simulations. The computation costs of the algorithm were reduced using sparse Polynomial Chaos Expansion (PCE) metamodels. The PCE metamodels are spectral variance decompositions of the original model with low training cost and wide coverage in the parameter domain [13]. For the chosen $\mathcal{M}(\xi)$, its PCE metamodel, $\mathrm{PCE_{EM}}(\xi)$, is developed from the (2), given below.

$$\mathcal{M}(\xi) \cong \mathrm{PCE_{EM}}(\xi) = \sum_{j=0}^{\infty} a_j \Psi_j(\xi) \cong \sum_{j=0}^{N} a_j \Psi_j(\xi). \tag{2}$$

In (2), $a_j \in \mathbb{R}$ is a set of scalar coefficients and the $\Psi_j(\xi) \in \mathbb{R}$ form a polynomial orthonormal basis [14]. For practical reasons and to avoid over fitting conditions, the metamodels were limited to $N$ (=20) expansions. In this study, the $\mathrm{PCE_{EM}}$ metamodel, $Y(\xi)$, was implemented the UQLab toolbox [15] with Legendre polynomial family with the uniform $[-1,+1]$ input distributions. Details of the metamodel implementation for the EM model can be found in [10].

*2.2. Probabilistic Particle Swarm Optimization*

The inversions of multi-dimensional EM models are ill-posed problems with the higher number of unknowns compared to the number of equations. For such problems, optimization algorithms are used to reach the optimum solution in the parameter space using different constraints. However, the optimization of multi-dimensional EM model inversions may result in multiple optimum solutions since different inputs can lead to similar outputs. The presence of multiple optimum solutions prevents the use of deterministic optimization algorithms that focus on a single solution. For the existence of multiple solutions, stochastic optimization algorithms can be considered as an alternative. In stochastic optimization, the procedure is initiated several times to obtain all local solutions in a given parameter space based on the defined set of rules.

In this study, the Particle Swarm Optimization (PSO) algorithm [16] is utilized, which is based on updating the position of the particles, i.e., possible solutions, until they converge to an optimum solution within a parameter space. In each iteration, the locations of the particles are updated according to the position of the particle with the best position. The iterations continue until the particles converge to a solution that agrees with the defined constraints. For the estimation of rice morphology parameters from copolar X-band SAR backscattering intensities, the $\mathrm{PCE_{EM}}$ metamodel is inverted with the stochastic PSO algorithm.

The fitness function for the PSO is given in (3) for HH and VV polarimetric channels and an arbitrary $i^{th}$ iteration. The fitness function is defined to minimize the difference between measured $\sigma^0_{\mathrm{HH,VV}}$ and estimated $\bar{\sigma}^0_{\mathrm{HH,VV}}$ values. The consistency of the solutions for different polarimetric channels is provided by considering the same input vector for both HH and VV polarimetric channels.

$$\min \mathbb{C}_i = (\sigma^0_{\mathrm{HH}} - \bar{\sigma}^0_{\mathrm{HH}_i})^2 + (\sigma^0_{\mathrm{VV}} - \bar{\sigma}^0_{\mathrm{VV}_i})^2 \quad i = 1 \dots K \tag{3}$$

Optimization problems need constraints to simplify the problem by reducing the dimensional complexity. In $\mathrm{PCE_{EM}}$ metamodel inversion, three plant morphology dependent constraints were established that are based on the natural limitations excerpted from the available rice morphology data.

**Positivity constraint** ensures positive and real morphological estimations for all iterations.

**Min-Max constraint** limits the morphological estimations based on the phenological phase boundaries. Phenological phases are defined by the International Rice Research Institute (IRRI) [17].

The scale divides the growth cycle to five major phases. Figure 3 presents the boundaries obtained from ground data, for the chosen morphological parameters.

**Natural limitations** provide non-linear relationships among rice morphology parameters during their development. These relations eliminates the solutions which might be impossible for a healthy plant, such as a 10-cm-tall stalk having two 50-cm-length leaves. This condition restricts the parameter space with a convex hull. Convex hull defines non-linear boundaries in a parameter space according to the experimental ground measurements that involve the agronomically possible solutions.

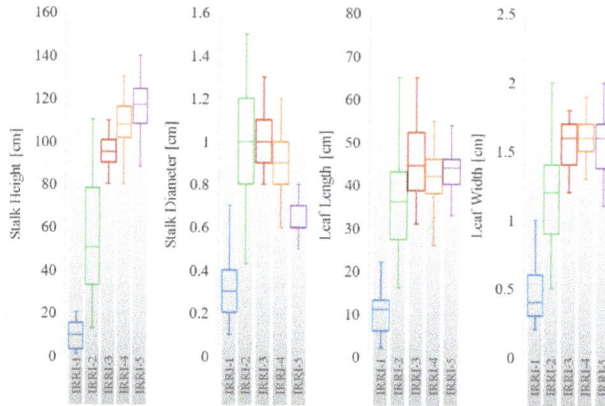

**Figure 3.** Temporal variations of the rice crop biophysical collected between 2013–2015 from the regularly conducted ground campaigns in Ipsala (Turkey). Due to technical limitations, stalk diameter data was not measured in 2015. The measurements are grouped according to the IRRI growth phases as a Box-and-Whisker plot. Box presents the information for the quartiles while the whiskers present minimum and maximum values for IRRI growth stages IRRI-1 [Early vegetative], IRRI-2 [Late vegetative], IRRI-3 [Early reproductive], IRRI-4 [Late reproductive], IRRI-5 [Maturative].

The stochastic PSO optimization provides distributions for each morphology parameter. For the accuracy analysis, the mean value of the resulting stochastic distributions is assigned as the estimated dimensions of the rice morphological parameters. For a single stochastic PSO optimization, a total number of 200 iterations was found to be sufficient for the optimization convergence which changes less than 0.1% regarding the mean of the estimated values.

## 3. Datasets

### 3.1. The Ipsala Test Site and Ground Data

The selected test area, Ipsala, is located in the North-West part of Turkey with its center at 37°7'53" N and 6°19'32" N coordinates. Ipsala is one of the biggest rice cultivation sites in Turkey with approximate acreage of 190 square kilometers. Based on the knowledge gathered from the Trakya Agriculture Research Institute (TARI), rice cultivation is done in the area by local farmers between May and September.

As shown in Figure 4, field campaigns were conducted ±5 days of SAR acquisitions to have representative rice morphology parameters. In each year, the test fields were selected by the expertise of the TARI researchers. To monitor the evolution of the morphology parameters, at each field following parameters were measured: above water stalk height, stalk diameter, leaf length, leaf width, the number of plants per m$^2$, the number of tillers per plant, and the number of leaves per tiller. In Figure 3 the evolution of some morphology parameters is depicted for each IRRI phase using a box-whisker plot.

For each parameter, a quasi-linear increase is observed until IRRI-3. From IRRI-3 on, a decrease is noticed in the stalk diameter due to reduced water content.

## 3.2. SAR Data

During the study, the Ipsala test site is monitored with the data acquired from the TerraSAR-X satellite with an average incidence angle of 31°. The TerraSAR-X satellite has the central frequency of 9.65 GHz and temporal resolution of 11 days. The acquisition dates are given in Figure 4. The data were delivered in single look complex format and were spatially and temporally co-registered by bilinear interpolation.

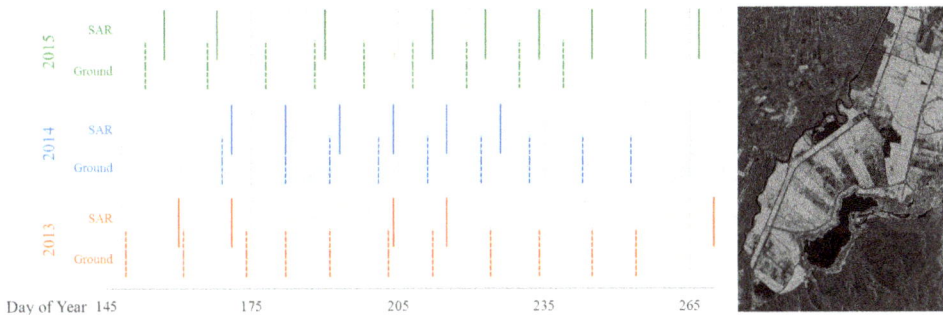

**Figure 4.** The ground and Synthetic Aperture Radar (SAR) data collection dates for the cultivation period between 2013 and 2015. On the right side, the VV channel PolSAR intensity image is also provided from the 30 June 2013. In the image, rice fields can be detected with their higher backscattering intensities.

## 4. Results and Discussion

In this section, we present the stochastic inversion results of the same $PCE_{EM}$ metamodel for the estimation of stalk height, leaf length and leaf width parameters from X-band co-polar SAR data on a dataset spanned over three cultivation periods. For the analysis, the noise in the TerraSAR-X data was reduced using 13 × 13 boxcar smoothing windows, resulting in 33 m × 25 m spatial resolution. The estimation accuracies of the chosen parameters are evaluated based on their correlation against the ground measured values. The results are reported with their coefficient of determination ($R^2$) and the root-mean-square error (RMSE) values.

Table 1 provides the input parameters for the $PCE_{EM}$ metamodel evolutions that were assumed constant. The details of the EM model, its metamodel, their simulation accuracies and growth phase based global sensitivity analysis results of the $PCE_{EM}$ metamodel are provided in [12].

**Table 1.** Input parameters that are assumed constant during the EM model and $PCE_{EM}$ metamodel simulations.

| Parameter | Value |
|---|---|
| Central frequency | 9.65 GHz |
| Dielectric constant ($\epsilon_{s,l}$) | 25 + 8 j |
| Dielectric constant ($\epsilon_g$) | 70 + 20 j |
| Average incidence angle ($\theta$) | 31° |
| Distance to target | 514 km |
| Illuminated area x-size | 2.58 m |
| Illuminated area y-size | 1.79 m |
| Number of MC iterations | 200 |

The chosen EM model and the proposed stochastic inversion approach consider the simplified plant morphology. As observed in Figure 3, the evolution of chosen morphological parameters are related to each other. In the proposed method these morphology based relations provide the base of the optimization constraints. As the stochastic inversion algorithm provides multi-dimensional parameter distributions, the mean values of the estimations are used for the calculation of $R^2$ and RMSE values.

The stochastic inversion results for stalk height, leaf length, and leaf width parameters are presented in Figure 5 and Table 2. The $R^2$ and RMSE were calculated against the estimated biophysical parameters are given for the cultivation periods from 2013 to 2015. The accuracy analyses were conducted using a total number of 93 different $\sigma^0$ values (32 from 2013, 25 from 2014, and 36 from 2015) measured from different fields having different morphologies, agricultural practices and environmental conditions.

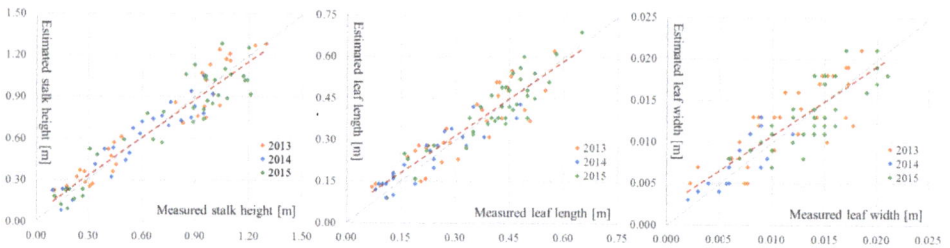

**Figure 5.** The measured values of stalk height, leaf length and leaf width parameters versus their estimations given in a correlation scatter plot for years 2013 to 2015 with a reference line. Symbols are colored with respect to the year: 2013 (•), 2014 (•) and 2015 (•).

The global sensitivity analysis of the $PCE_{EM}$ metamodel was previously discussed in [12]. The analysis results emphasize the importance of stalk height, leaf length and structural density parameters (number of plant components in a unit area) throughout the growth cycle. The following deductions can be made from the stochastic inversion results.

**Table 2.** Accuracy analysis of the stochastic inversion of $PCE_{EM}$ metamodel given with the calculated $R^2$ and root-mean-square error (RMSE) values for the years 2013, 2014, 2015 and the complete dataset.

|  | Stalk Height | | Leaf Length | | Leaf Width | |
|---|---|---|---|---|---|---|
|  | $R^2$ | RMSE | $R^2$ | RMSE | $R^2$ | RMSE |
| 2013 | 0.921 | 0.108 | 0.804 | 0.069 | 0.646 | 0.003 |
| 2014 | 0.891 | 0.092 | 0.831 | 0.048 | 0.628 | 0.002 |
| 2015 | 0.868 | 0.135 | 0.776 | 0.055 | 0.613 | 0.003 |
| Complete | 0.894 | 0.116 | 0.848 | 0.057 | 0.717 | 0.003 |

**Stalk Height** estimations had the highest accuracy ($R^2 \geq 0.86$) considering the entire dataset of three cultivation periods. As presented in Figure 5, the stalk height estimation results are slightly over-estimated for rice canopies shorter than 0.6 m and under-estimated for the taller canopies. The performance of the algorithm for the 2015 dataset was calculated to be lower compared to the other years with an RMSE value of 13.5 cm. This situation can be related to the variance of the $\sigma^0$ values and the $PCE_{EM}$ metamodel simulations for the corresponding morphology parameter ranges in the parameter space. On the other hand, the estimation bias between the measured and estimated values shows an increasing spread with increasing stalk height. The variation in the bias can be interpreted by the presence of plants with varying physical structures and similar backscattering behaviors at later phases of the growth cycle. The RMSE values were calculated to be less than 13.5 cm, which are acceptable for the stalk height estimations calculated from copolar SAR data.

**Leaf Length** estimation accuracy is calculated to be lower than the stalk height estimation accuracy for the dataset of three cultivation periods. As shown in Figure 5, the leaf lengths are mainly over-estimated when they are shorter than 50 cm and under-estimated when they become longer. The stochastic inversion results of the $PCE_{EM}$ metamodel shows highly accurate results ($R^2 \geq 0.78$ and RMSE $\leq 7$ cm) for the evaluations from the complete dataset. From three different years, the lowest accuracy was obtained from the data acquired during 2015, as it was observed for the stalk height. Regarding the estimation bias, it is noted that the errors tend to increase with increasing leaf length. Similar to the interpretation provided for stalk height estimations, the increasing error with increasing leaf length can be explained by the presence of higher variance in the parameter space at later growth phases.

**Leaf Width** is morphologically related to the leaf length due to natural growth limitations. The accuracy analysis on leaf width estimations presented acceptable values with $R^2$ values higher than 0.61 and RMSE values lower than 3 mm. Similar to stalk height and leaf length parameters, the lowest accuracy was again obtained from the copolar SAR measurements of 2015. The estimation bias of the analysis is observed to vary between ±5 mm for the complete growth cycles of three years.

The stochastic inversion of the $PCE_{EM}$ metamodel provided successful estimations for stalk height and leaf length parameters for the cultivation periods of three years. However, from the chosen rice morphology parameters, leaf width estimations had lower accuracies with higher relative RMSE values. This situation is supported by the global sensitivity analysis results, which states the importance of stalk height and leaf length parameters on the $PCE_{EM}$ metamodel simulations, while mentions the lower effect of the leaf width [12].

The accuracy analysis exhibited in this study combines 93 different rice plant morphology and SAR measurements collected from the cultivation periods of 2013 to 2015. Concerning the presence of various environmental factors and agricultural practices, the results are considered to be representative for the wet-cultivated and broadcast seeded rice fields that are located in Ipsala (Turkey). Therefore, the results are encouraging for the development of new management practices, which can use the estimated morphology parameters to interpret the yield and the detailed growth stage.

## 5. Overview and Summary

In this research, a stochastic inversion method has been presented to invert a multi-dimensional $PCE_{EM}$ metamodel, trained from an EM model [11]. Apart from the previous studies that focuses on the estimation of the stalk height, this study extends the estimations to the stalk height, leaf length and leaf width parameters of rice plants by considering the natural growth limitations. This study was conducted over three cultivation cycles for validation purposes.

We have tested the proposed inversion algorithm on a three-year cultivation period of ground, and copolar high spatial resolution and high frequency SAR datasets. The data were acquired by TerraSAR-X over broadcast seeded and flooded rice fields. For method validation we trained the $PCE_{EM}$ metamodel by employing the EM model [11], ground measurements and field average SAR backscattering intensities.

As a result, we obtained significant correlations between the estimated and measured values of stalk height, leaf length and leaf width parameters using the proposed $PCE_{EM}$ metamodel inversion scheme. From the analysis, we calculated RMSE values less than 13.5 cm for stalk height, 6.9 cm for leaf length and 34 mm for leaf width parameters. The results pointed out that the use of spaceborne X-band PolSAR data is powerful for the development of new agriculture monitoring practices. We should mention that the overall performance of the proposed approach mainly relies on the accuracy of the EM model and the $PCE_{EM}$ metamodel, which shows the importance of EM model selection.

The presented algorithm has presented several advantages such as coupling the agronomical growth rules with the EM models for efficiency, inversion with stochastic optimization to handle environmental variability and most importantly being computationally efficient with the use of metamodels that are trained for once. In the presented study, the requirement of a priori knowledge of the growth phase and plant morphology information can be seen as a disadvantage. However, considering the importance of rice as the main source or food and income for several highly populated countries, related information can be found in the crop databases. Besides, the literature currently covers several different methodologies to determine the growth phase of rice plants [3,18,19].

In the future, our studies are going forward to improve the approach by substituting the *a priori* growth phase information with canopy height estimations that can be directly calculated from Pol-InSAR data. In addition, it is planned to extend the applicability of the inversion procedure to monitor other major crops.

**Acknowledgments:** This work has been supported by the Scientific and Technological Research Council of Turkey (TUBITAK) under Project 113Y446 and by the German Aerospace Center (DLR) under project XTILAND1476. Besides, the authors would like to thank for the support of Dr. Stefano Marelli and Prof. Dr. Bruno Sudret from the Chair of Risk, Safety & Uncertainty Quantification from ETH Zurich in setting up and providing help with the Polynomial Chaos Expansion metamodel and to the Directorate of Trakya Agricultural Research Institute with the ground campaigns.

**Author Contributions:** O.Y. developed and implemented the proposed inversion methodology, interpreted the results, wrote and revised the paper; E.E. supervised the work, contributed to the underlying ideas as well as to the interpretation and discussion of the results and revised the paper; I.H. supervised the work, contributed to the underlying ideas and revised the paper.

**Conflicts of Interest:** The authors declare no conflict of interest.

## References

1. Mulla, D.J. Twenty five years of remote sensing in precision agriculture: Key advances and remaining knowledge gaps. *Biosyst. Eng.* **2013**, *114*, 358–371.
2. Inoue, Y.; Kurosu, T.; Maeno, H.; Uratsuka, S.; Kozu, T.; Dabrowska-Zielinska, K.; Qi, J. Season-long daily measurements of multifrequency (Ka, Ku, X, C, and L) and full-polarization backscatter signatures over paddy rice field and their relationship with biological variables. *Remote Sens. Environ.* **2002**, *81*, 194–204.
3. Lopez-Sanchez, J.M.; Cloude, S.R.; Ballester-Berman, J.D. Rice Phenology Monitoring by Means of SAR Polarimetry at X-Band. *IEEE Trans. Geosci. Remote Sens.* **2012**, *50*, 2695–2709.
4. Erten, E.; Rossi, C.; Yuzugullu, O. Polarization Impact in TanDEM-X Data over Vertical-Oriented Vegetation: The Paddy-Rice Case Study. *IEEE Geosci. Remote Sens. Lett.* **2015**, *12*, 1501–1505.
5. Erten, E.; Lopez-Sanchez, J.M.; Yuzugullu, O.; Hajnsek, I. Retrieval of agricultural crop height from space: A comparison of SAR techniques. *Remote Sens. Environ.* **2017**, *187*, 130–144.
6. Ballester-Berman, J.; Lopez-Sanchez, J.; Fortuny-Guasch, J. Retrieval of biophysical parameters of agricultural crops using polarimetric SAR interferometry. *IEEE Trans. Geosci. Remote Sens.* **2005**, *43*, 683–694.
7. Lopez-Sanchez, J.M.; Vicente-Guijalba, F.; Erten, E.; Campos-Taberner, M.; Garcia-Haro, F.J. Retrieval of vegetation height in rice fields using polarimetric SAR interferometry with TanDEM-X data. *Remote Sens. Environ.* **2017**, *192*, 30–44.
8. Zhang, Y.; Liu, X.; Su, S.; Wang, C. Retrieving canopy height and density of paddy rice from Radarsat-2 images with a canopy scattering model. *Int. J. Appl. Earth Obs. Geoinf.* **2014**, *28*, 170–180.
9. Yuzugullu, O.; Erten, E.; Hajnsek, I. Estimation of Rice Crop Height From X- and C-Band PolSAR by Metamodel-Based Optimization. *IEEE J. Sel. Top. Appl. Earth Obs. Remote Sens.* **2017**, *10*, 194–204.
10. Yuzugullu, O.; Marelli, S.; Erten, E.; Sudret, B.; Hajnsek, I. Determining Rice Growth Stage with X-Band SAR: A Metamodel Based Inversion. *Remote Sens.* **2017**, *9*, 460.
11. Toan, T.L.; Ribbes, F.; Wang, L.F.; Floury, N.; Ding, K.H.; Kong, J.A.; Fujita, M.; Kurosu, T. Rice crop mapping and monitoring using ERS-1 data based on experiment and modeling results. *IEEE Trans. Geosci. Remote Sens.* **1997**, *35*, 41–56.
12. Yuzugullu, O.; Marelli, S.; Erten, E.; Sudret, B.; Hajnsek, I. Global sensitivity analysis of a morphology based electromagnetic scattering model. In Proceedings of the 2015 IEEE International Geoscience and Remote Sensing Symposium (IGARSS), Milan, Italy, 26–31 July 2015; Institute of Electrical and Electronics Engineers (IEEE): Piscataway, NJ, USA, 2015.
13. Blatman, G.; Sudret, B. Efficient computation of global sensitivity indices using sparse polynomial chaos expansions. *Reliab. Eng. Syst. Saf.* **2010**, *95*, 1216–1229.

14. Sudret, B. Global sensitivity analysis using polynomial chaos expansions. *Reliab. Eng. Syst. Saf.* **2008**, *93*, 964–979.

15. Marelli, S.; Sudret, B. *UQLab User Manual-Polynomial Chaos Expansion*; Technical Report, Report-UQLab-V0.9-104; Chair of Risk, Safety & Uncertainty Quantification, ETH Zurich: Zurich, Switzerland, 2015.

16. Eberhart, R.; Kennedy, J. A new optimizer using particle swarm theory. In Proceedings of the Sixth International Symposium on Micro Machine and Human Science, Nagoya City, Japan, 4–6 October 1995; Institute of Electrical and Electronics Engineers (IEEE): Piscataway, NJ, USA, 1995.

17. GRiSP (Global Rice Science Partnership). *Rice Almanac*, 4th ed.; International Rice Research Institute: Los Banos, Philippines, 2011.

18. Vicente-Guijalba, F.; Martinez-Marin, T.; Lopez-Sanchez, J.M. Crop Phenology Estimation Using a Multitemporal Model and a Kalman Filtering Strategy. *IEEE Geosci. Remote Sens. Lett.* **2014**, *11*, 1081–1085.

19. Yuzugullu, O.; Erten, E.; Hajnsek, I. Rice Growth Monitoring by Means of X-Band Co-polar SAR: Feature Clustering and BBCH Scale. *IEEE Geosci. Remote Sens. Lett.* **2015**, *12*, 1218–1222.

*applied sciences*

MDPI

Article

# Scattering Characteristics of X-, C- and L-Band PolSAR Data Examined for the Tundra Environment of the Tuktoyaktuk Peninsula, Canada

Tobias Ullmann [1],*, Sarah N. Banks [2], Andreas Schmitt [3] and Thomas Jagdhuber [4]

[1]  Institute Geography and Geology, University of Wuerzburg, D-97074 Wuerzburg, Germany
[2]  National Wildlife Research Center (NWRC), Carleton University, Ottawa, ON K1A 0H3, Canada; sbanks@connect.carleton.da
[3]  Department of Geoinformatics, University of Applied Sciences Munich, D-80333 Munich, Germany; andreas.schmitt@hm.edu
[4]  German Aerospace Center (DLR), Microwaves and Radar Institute, D-82234 Wessling, Germany; thomas.jagdhuber@dlr.de
*  Correspondence: tobias.ullmann@uni-wuerzburg.de; Tel.: +49-931-31-86865

Academic Editor: Carlos López-Martínez
Received: 27 April 2017; Accepted: 3 June 2017; Published: 8 June 2017

**Abstract:** In this study, polarimetric Synthetic Aperture Radar (PolSAR) data at X-, C- and L-Bands, acquired by the satellites: TerraSAR-X (2011), Radarsat-2 (2011), ALOS (2010) and ALOS-2 (2016), were used to characterize the tundra land cover of a test site located close to the town of Tuktoyaktuk, NWT, Canada. Using available in situ ground data collected in 2010 and 2012, we investigate PolSAR scattering characteristics of common tundra land cover classes at X-, C- and L-Bands. Several decomposition features of quad-, co-, and cross-polarized data were compared, the correlation between them was investigated, and the class separability offered by their different feature spaces was analyzed. Certain PolSAR features at each wavelength were sensitive to the land cover and exhibited distinct scattering characteristics. Use of shorter wavelength imagery (X and C) was beneficial for the characterization of wetland and tundra vegetation, while L-Band data highlighted differences of the bare ground classes better. The Kennaugh Matrix decomposition applied in this study provided a unified framework to store, process, and analyze all data consistently, and the matrix offered a favorable feature space for class separation. Of all elements of the quad-polarized Kennaugh Matrix, the intensity based elements K0, K1, K2, K3 and K4 were found to be most valuable for class discrimination. These elements contributed to better class separation as indicated by an increase of the separability metrics squared Jefferys Matusita Distance and Transformed Divergence. The increase in separability was up to 57% for Radarsat-2 and up to 18% for ALOS-2 data.

**Keywords:** PolSAR; dual polarimetry; quad polarimetry; decomposition; TerraSAR-X; Radarsat-2; ALOS; ALOS-2; tundra; arctic

## 1. Introduction

Polarimetric Synthetic Aperture Radar (PolSAR) data from an increasing number of different satellite systems has become available—or will become available in the near future—for up-to-date Earth observation and environmental monitoring. Microwave data, e.g., acquired by Sentinel-1, ALOS-2, or in the future, by the RADARSAT Constellation Mission, are capable of delivering remote sensing data at high spatial (<10 m) and temporal resolutions (<10 days); independent of weather and illumination conditions. Therefore they are well suited for characterizing and monitoring the dynamic nature of the land surface, especially in vast and remote regions like the Arctic. In light of the challenges associated with a changing climate and environment, such investigations are crucial for assessing and comprehending temporal and spatial changes. Several studies have therefore investigated the use of SAR, PolSAR, and SAR interferometry (InSAR) for characterizing Arctic land surfaces, particularly that of tundra environments.

Table 1 provides an overview of select studies that involved the use of active microwave imaging of Arctic tundra environments. As can be observed, shortwave C- and X-Band data were frequently used for the characterization of land cover and shoreline types, and most studies incorporated analysis of polarimetric information to relate observed values with certain types or states of the land surface [1–16]. C-Band and L-Band data stacks have been used primarily to determine surface movements in permafrost regions using interferometry [17–21]. This is likely driven by the opening of the ALOS archive and the capacity of the L-Band microwaves to penetrate the relatively small tundra vegetation, typically resulting in higher interferometric coherence, and leading to a more reliable estimate of surface movements. The seasonal thawing and freezing of the active layer is also a promising area for InSAR applications, and estimation of the active layer thickness/variations across the entirety of the Arctic is of particular importance considering recent warming trends [22–24].

**Table 1.** Select studies that employed Synthetic Aperture Radar (SAR), polarimetric SAR (PolSAR) and InSAR (SAR interferometry) data and methods for the characterization of tundra (permafrost) landscapes and their dynamics; ERS: European Remote Sensing Satellite; R-1: Radarsat-1; TSX: TerraSAR-X; R-2: Radarsat-2; ALOS: Advanced Land Observing Satellite.

| Study Area | Sensor | Task and Method | Year and Reference |
|---|---|---|---|
| Alaska, USA | ERS | Bathymetric mapping of shallow water via time series Analysis | 1996 & 2000 [1,2] |
| Mackenzie Delta Region, Canada | ERS | Delineation of delta ecozones via InSAR—Coherence | 2001 [3] |
| Nova Scotia, Canada | R-1 | Mapping of geomorphological units in the intertidal zone via unsupervised classification | 2001 [4] |
| Quebec, Canada | TSX | Monitoring of permafrost dynamics via InSAR | 2011 [5] |
| Herschel Island, Canada | TSX R-2 ALOS | Monitoring of surface movements via InSAR | 2009 & 2011 [16,17] |
| Mackenzie Delta Region, Canada | R-2 | Classification of tundra land cover and shoreline types via PolSAR | 2011 & 2014 [6,7] |
| Lena Delta, Russia | TSX R-2 | Characterization of melt onset and geomorphological units via PolSAR | 2012 [8] |
| Alaska, USA | TSX | Characterization of post-drainage succession via time series analysis and PolSAR | 2012 [9] |
| Sodankylä, Finland | R-2 | Identification of soil freezing and thawing states | 2014 [10] |
| Richards Island, Canada | TSX R-2 | Classification of tundra land cover via PolSAR | 2014 [11] |
| Baffin Island, Canada | R-2 | Monitoring of surface movements via InSAR | 2014 [18] |
| Northern Canada | R-2 | Modeling of phytomass via PolSAR | 2014 [12] |
| Dease Strait, Nunavut, Canada | R-2 | Classification of shoreline types via PolSAR | 2015 [13] |
| Barrow, Alaska | ALOS | Active-layer thickness estimation via InSAR | 2015 [19] |
| Mackenzie Delta Region, Canada | TSX R-2 | Characterization of tundra land cover via PolSAR | 2016 [14] |
| Northern Alaska, USA | ALOS | Active-layer change and subsidence monitoring via InSAR | 2016 [20] |
| Northern Qinghai-Tibetan Plateau | ALOS | Active-layer change via InSAR | 2017 [21] |
| Yamal Peninsula, Russia | TSX | Active-layer thickness estimation via backscatter intensity | 2017 [15] |

Most of the studies identified in Table 1, employed data of a single sensor and only few involved multi-frequency SAR/PolSAR/InSAR information, e.g., dealing with some combination of X-, C- or L-Band data [8,11,14,17]. Among the selected studies, the quad-polarization mode of Radarsat-2 was most frequently employed, followed by the dual HH/HV-polarization mode of ALOS and

TerraSAR-X, and the dual HH/VV-polarization mode of TerraSAR-X. For these different datasets, several polarimetric decomposition approaches were applied, including: the Eigen-decomposition (with the features Entropy, Anisotropy and Alpha Scattering Angle) [25,26], the Yamaguchi Decomposition [27], the Freeman-Durden Decomposition [28], and the Touzi Decomposition [29]. Two decomposition models for HH/VV-polarized data were further proposed by [30] and recently by [14].

In light of this previous research, and our preliminary investigations [6,7,11,13,14], we analyze X-, C- and L-Band PolSAR data in order to characterize scattering properties of select tundra land cover classes for a test site in the Arctic. In addition, this study incorporates quad-polarized data of ALOS-2, a novel compilation of in situ data for the test site, and a complete utilization of the Kennaugh Matrix approach, recently presented by [31]. The Kennaugh Matrix approach offers a unified framework for processing polarimetric information of different polarization modes (quad-, dual- and compact-polarized data). It can be used to represent targets both incoherently and coherently, and can be converted into all of the well-established decomposition models, for all wavelengths [25]. Information on the Kennaugh Matrix framework is provided in the subsequent Section 2.2.2 Polarimetric SAR Data and Decompositions of this manuscript and in Appendix A.

The objectives of this research are therefore: First, process and analyze decomposition features of quad- and dual-polarized data of different sensors at three different frequencies. Second, investigate the backscattering of generalized tundra land cover classes for quad- and dual-polarized data of X-, C- and L-Band data. Third, investigate the correlation among PolSAR features of quad- and dual-polarized decomposition techniques. Fourth, benchmark and rank all PolSAR (decomposition) features in terms of class separability, and identify feature spaces and parameters that are most meaningful for characterizing the tundra land cover.

This manuscript is structured as follows: The subsequent section provides details on the materials and methods, as well as information on the location and environment of the test site selected for this research. Further, this section describes and lists the available land cover reference information, and the PolSAR data, including: quad- and dual-polarized data of Radarsat-2 (R-2), TerraSAR-X (TSX), ALOS and ALOS-2. Subsequently, the data processing and all polarimetric decompositions applied to the data are described, as well as separability measures/metrics used to analyse them, including: Transformed Divergence and Jefferys Matusita Distance.

The third section presents the results of the correlation analyses of C-Band R-2 and L-Band ALOS-2 data. Backscatter characteristics and select decomposition features of the land cover classes are presented and analysed via box- and scatterplots for X-, C- and L-Band data. Afterward, the results of the separability analysis and the feature selection are presented. Section four discusses the main findings, while section five provides a summary of the study; major conclusions are drawn and an outlook on future work is given.

## 2. Materials and Methods

### 2.1. Test Site Description

The study area is located at the northern extent of the mainland of the Northwest Territories, Canada (see Figure 1a). The region is part of the Canadian Arctic and lies entirely north of the tree-line along the coastal tundra plains of the Southern Arctic Ecozone [32]. The climate here is characterized cold winters, followed by short and cool summers. The mean annual air temperature at the climate station Tuktoyaktuk is $-1\,°C$ (1971–2000), and the mean air temperature between October and April is below $-10\,°C$. The average precipitation is about 150 mm [33]. The ground surface is characterized by the presence of continuous permafrost and its thickness is estimated to be up to 600 m [34]. Therefore, the soils of the Tuktoyaktuk Peninsula are Cryosols, with an active layer thickness of several centimeters to decimeters. The soils developed on glacial deposits of Pleistocene to Holocene age [32], and current morphodynamics are dominated by periglacial processes. The land surface is therefore characterized by low-lying and flat coastal plains, rolling hills, thermokarst lakes and pingos, and extensive networks of high- and low-centered ice wedges [33]. The ground surface is also characterized by tundra vegetation, with upland tundra usually composed of short herbaceous vegetation and shrubs (dwarf shrubs up to tall shrubs). The wetland vegetation (grasses, sedges or rushes) is frequently at or near water bodies, e.g., at drained lakes, or in the flat and low-lying intertidal zone. Depending on the coastal currents, the beach zone is characterized by fine sandy material, mixed sediments dominated by gravel, pebble or cobble and driftwood accumulations [33].

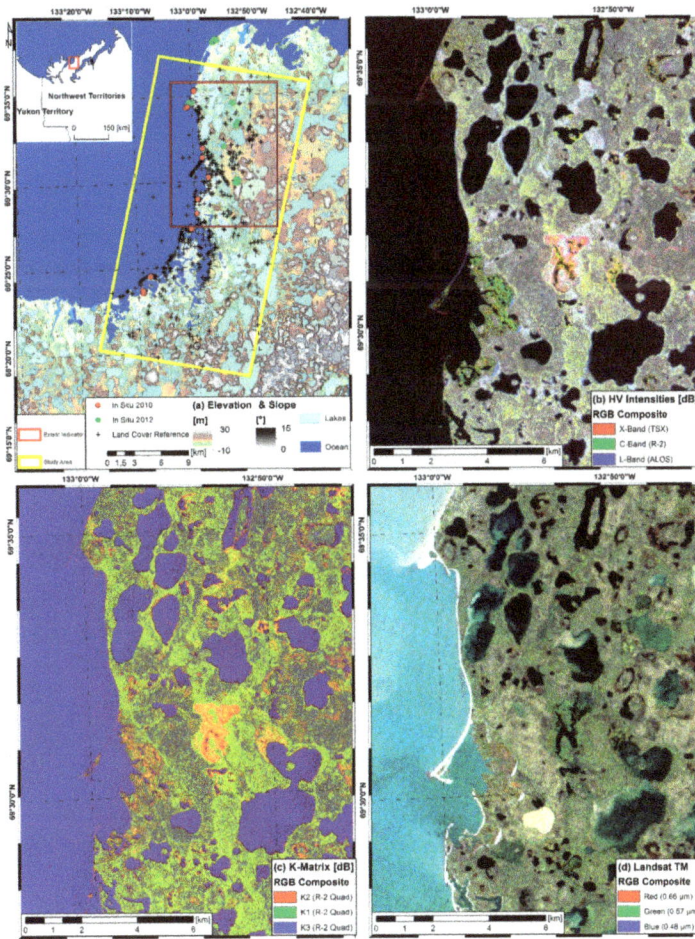

**Figure 1.** Location of the test site: Tuktoyaktuk (Northwest Territories, Canada) and RGB composites of remote sensing imagery: (**a**) elevation and slope of intermediate TanDEM-X DEM, coverage of TerraSAR-X, Radasat-2, ALOS and ALOS-2 imagery, locations of in situ field work in 2010 and 2012, locations of land cover reference samples derived from high resolution ortho-photos, extent indicator of the subsequent sub-figures (red rectangle); (**b**) RGB false-color composite of sigma nought HV intensities [dB] of X-Band (TerraSAR-X, 2011), C-Band (Radarsat-2, 2011) and L-Band (ALOS, 2010); (**c**) RGB false-color composite of Kennaugh Matrix Elements K2, K1 and K3 of quad-polarized Radarsat-2 data (see Section 2.2.2 Polarimetric SAR Data and Decompositions for more details on the Kennaugh Matrix Elements); (**d**) Landsat TM (2011) true-color composite of red, green and blue surface reflectance. RGB channels are stretched linearly between 1 and 99% of the data range.

## 2.2. Database

### 2.2.1. Land Cover Reference

In situ data on the land cover of the test site was acquired during two ground truth campaigns in the summer months of 2010 and 2012. The field work was organized and conducted by Carleton University, (Ottawa, ON, Canada), the NWRC (National Wildlife Research Centre, Ottawa, ON, Canada), and the University of Wuerzburg (Institute of Geography and Geology, Wuerzburg, Germany). Combining the land cover information of our preliminary studies [6,7,11,14], the land cover reference was merged to a common database showing the eight land cover classes listed and described in Table 2. During the field campaigns information on the land cover of representative homogenous locations was recorded, categorized and mapped. The in situ classification of the land cover was completed by field experts, and within the frameworks of [35,36]. As specified in Table 2 the cut off criteria for the separation of Shrub (ST) and Herb Dominated Tundra (HT) was the presence of dwarf shrubs with a height greater or less than 0.25 m. The separation between Sand (BS) and Mixed Sediment (BM) was based on the presence of pebble, which had to cover >50% of the surface; the bare ground samples were considered homogeneous if 10% or less were "other" materials or vegetation. Locations were classified as Driftwood Accumulation (BW) if more than 80% of the ground were cover by driftwood. Wetland Vegetation Communities (WT) were dominated by grasses, sedges or rushes and frequently occurred inland at creeks and drained lakes. The locations classified as Inundated Low Lying Tundra (WI) exhibited tundra vegetation communities at or near a water body and were most prominent in the low-lying coastal supratidal north of the town of Tuktoyaktuk.

**Table 2.** Land cover classes considered in the analysis; description, class abbreviations (Abbr.), and class color coding. Bare Ground samples were considered homogeneous if 10% or less were "other" materials or vegetation. The letters "W", "B", "T" of the class abbreviations refer to *Wetland*, *Bare Ground*, and *Tundra* land cover classes.

| Land Cover Class Name | | Description | Abbr. | Class Color |
|---|---|---|---|---|
| Tundra Vegetation "T" | Herb Dominated Tundra | upland tundra composed of short herbaceous vegetation and low shrubs (<25 cm) | HT | |
| | Shrub Dominated Tundra | upland tundra dominated by tall shrubs (>25 cm) | ST | |
| Bare Ground "B" | Sand | sediment dominated by sand (0.0625–2.0 mm) | BS | |
| | Mixed Sediment | mixed sandy sediment dominated by gravel, pebble or cobble (2.0–256.0 mm) and without woody debris | BM | |
| | Driftwood Accumulation | accumulations of driftwood (>80%) | BW | |
| Wetland "W" | Wetland | wetland vegetation communities dominated by grasses, sedges or rushes | WT | |
| | Inundated Low Lying Tundra | vegetated tundra at or near a water body | WI | |
| Water | Permanent Water Bodies | ocean, inland lakes, river channels and ponds | OL | |

Further, Figure 2 provides example photographs of select land cover classes. In total, information from more than fifty ground truth sites were available. Additionally, the number of samples was increased using high resolution airborne imagery with less than one meter spatial resolution provided by [34]. The generation of the land cover reference database was completed and locations of homogenous land cover were digitized using the airborne imagery. The reference information was then available in polygon format. Afterward, random sampling was applied, in order to generate 200 samples for each of the land cover classes listed in Table 2. Note that each individual point was selected to represent homogenous information for a certain land cover class (i.e., areas of mixed land covers were avoided). Figure 1a shows the locations of some sites visited and indicates the centers of the manually digitized polygons that exhibited homogeneous land coverage in the airborne imagery.

**Figure 2.** Example in situ imagery of select land cover classes of interest: (**a**) sand (BS); (**b**) mixed sediment (BM); (**c**) driftwood accumulation (BW); (**d**) herb dominated tundra (HT); (**e**) shrub dominated tundra (ST) and (**f**) wetland (WT). Photos were taken in 2012 on the Tuktoyaktuk Peninsula by Tobias Ullmann.

Additionally, a second set of 50,000 land samples was randomly generated, representing approximately 10% of all land pixels inside the common coverage of TSX, ALOS, ALOS-2 and R-2. This second set was generated independent of the land cover and was used to estimate the correlations among the PolSAR features. It is assumed that this sample represents the natural distribution of the relevant land cover classes.

2.2.2. Polarimetric SAR Data and Decompositions

PolSAR data from R-2, TSX, ALOS and ALOS-2 were available for the test site. Table 3a lists the main acquisition parameters and shows that all data was acquired in the summer months, during the growing season of the tundra vegetation. Note that the ALOS-2 data was acquired at a steep incidence angle (28°) in 2016, while the data of the other sensors were acquired with incidence angles between 34° and 40° in 2010/2011; along with the in situ reference data. Changes of the land cover were considered to be of less relevance for the analysis, considering the recent studies on the decadal changes in composition of the tundra vegetation here [37,38] the spatial resolution of the data, and the rather broadly defined classes. Figure 1b shows a false-color composite of the HV intensities of TerraSAR-X (TSX), R-2 and ALOS.

Table 3. (a) Acquisition parameters of TerraSAR-X (TSX), Radarsat-2 (R-2), ALOS PALSAR (ALOS), and ALOS-2 PALSAR-2 (ALOS-2); (b) acquisition parameters of Landsat TM imagery and (c) overview on the polarimetric features considered in the analysis.

**(a) PolSAR Database**

| Sensor | Wavelength/Band | Date of Acquisition | Mode | Polarization | Incidence Angle |
|---|---|---|---|---|---|
| TSX | 3.1 cm/X | 3 August 2011 | Stripmap | Dual HH/VV | 38.8° |
| TSX | 3.1 cm/X | 23 July 2011 | Stripmap | Dual HH/HV | 38.8° |
| R-2 | 5.5 cm/C | 19 August 2011 | Fine | Quad HH/HV/VH/VV | 40.5° |
| ALOS | 23.6 cm/L | 21 July 2010 | Fine Beam Dual (FBD) | Dual HH/HV | 34.3° |
| ALOS-2 | 24.2 cm/L | 15 September 2016 | Stripmap (SM) | Quad HH/HV/VH/VV | 28.4° |

**(b) Multispectral Imagery**

| Sensor | Spectrum | Date of Acquisition | Path/Row | Sun Azimuth/Elevation |
|---|---|---|---|---|
| Landsat TM 5 | 0.07 µm–0.27 µm (excluding the thermal band) | 19 August 2011 | 63/11 | 171°/32° |

**(c) Polarimetric Features**

| Name/Model | Polarization | Feature Name(s) | Feature Symbol(s) | Source |
|---|---|---|---|---|
| Polarimetric Channels (sigma nought intensities) | Single/Dual/Quad | n/a | HH<br>HV<br>VH<br>VV | n/a |
| Two Component Decomposition | Dual (HH/VV) | Double Bounce<br>Surface Scattering | DBL2<br>ODD2 | [14] |
| Two Component Decomposition | Dual (HH/VV) | Volume Scattering<br>Ground Scattering | VOL2<br>GRD2 | [30] |
| Yamaguchi Decomposition | Quad | Double Bounce<br>Volume Scattering<br>Surface Scattering | DBL3<br>VOL3<br>ODD3 | [27] |
| Eigen-decomposition/ Entropy/Alpha | Dual/Quad | Entropy<br>Alpha of T-Matrix<br>Alpha of C-Matrix | ENT<br>ALPT<br>ALPC | [25,26] |
| Kennaugh Matrix | Single/Dual/Quad | Kennaugh Matrix Elements; total intensity (K0), absorption elements (K1,K2,K3), diattenuation elements (K4,K5,K6), retardance elements (K7,K8,K9) | K0, K1, K2, K3, K4, K5, K6, K7, K8, K9 | [31] |

The following processing steps were applied to the PolSAR data: First, synthetic dual-polarized data (HH/VV, VV/VH and HH/HV) were generated from the R-2 and the ALOS-2 quad-polarized data. These synthetic dual-polarized datasets were thus not affected by temporal variations and showed identical speckle characteristics (compared to the quad-polarized data from which the subsets were taken), as such, this allowed for direct comparison of class separability as a function of polarization diversity, as opposed to differences in moisture, and plant phenology. Second, the Sinclair scattering matrices of all dual- and quad-polarized data of all wavelengths were converted to the corresponding Kennaugh Matrices [31]. Third, the data were multi-looked (minimum of four looks) in order to generate pixels with square ground range resolution. Forth, a simple boxcar filter with a window size of $3 \times 3$ pixels was applied. Fifth, the data were terrain corrected and geocoded using the Range-Doppler Approach [39]. All data were transformed to UTM WGS1984 Zone 8 coordinate system with 12 m spatial resolution using the TanDEM-X intermediate digital elevation model (DEM) and the projected local incidence angle derived from this DEM [40]. The data were processed as sigma nought intensities.

All of the preceding steps were completed in SNAP 5.0 (Sentinel Application Platform) released by the European Space Agency (ESA), Paris, France. The terrain corrected Kennaugh Matrices were then used to generate the polarimetric channels, the Yamaguchi Decomposition [27], the Eigen-decomposition with the features Entropy/Alpha/Anisotropy, the Two Component Ground-Volume Decomposition of [30] and the Two Component Surface-Diherdal Decomposition of [14] using IDL 8.5 and ENVI 5.3. All intensity features were scaled to decibels [dB]. The above mentioned decompositions are explained in more detail in the subsequent paragraphs.

Kennaugh Matrix—For quad-polarized data the Kennaugh Matrix (Mueller Matrix, respectively [41]) describes the relation between the radiated and received wave as a symmetric $4 \times 4$ matrix using ten real elements (K0–K9). It is the linear transformation of the four-dimensional Stokes vector ([42] p. 43 ff. and p. 83 ff.) in the backscatter-alignment coordinate system. Unlike the Covariance or the Coherency Matrix, the Kennaugh Matrix can describe both coherent and incoherent targets [42,43]. The elements K0, K1 and K2 are intensity-based elements, while K3 and K4 are based on the cross-polarized intensity and the co-polarized phase information. The elements K5, K6, K7, K8 and K9 are phase-only elements that tend to provide unique information from natural targets. All elements of the full Kennaugh Matrix can be grouped as follows [31]: First, the total intensity (K0); second, the absorption elements that describe the loss of polarization during the scattering process (K1, K2, K3); third, diattenuation elements that describe the change of the relation between two amplitude values during reflection (K4, K5, K6); fourth, retardance elements that describe the phase delay during scattering in a particular direction (K7, K8, K9). The definition of the Kennaugh Matrix and its elements for quad-polarized (A1), HH/VV-polarized (A2) and HH/HV- or VV/VH-polarized data (A3) are shown in Appendix A in accordance to [31]. The Kennaugh Matrix elements are linear combinations of the Coherency Matrix and combinations of K0, K1, K2 and K3 describe the diagonal elements of the Coherency Matrix ($T_{11}$, $T_{22}$, $T_{33}$), while combinations of K4, K5, K6, K7, K8 and K9 describe off-diagonal elements of the Coherency Matrix ($T_{12}$, $T_{13}$, $T_{21}$, $T_{23}$, $T_{31}$, $T_{32}$) [42]. The conversions of the Kennaugh Matrix to $3 \times 3$ Coherency Matrix ($\mathbf{T}$) of quad-polarized (A4) and $2 \times 2$ $\mathbf{T}$ of HH/VV-polarized data (A5) are shown in Appendix A in accordance to [31]. To generate all Kennaugh matrix elements requires quadrature polarized data, thus only a portion of can be generated using dual polarized data. For HH/VV-polarized data the Kennaugh Matrix consists of the elements K0, K3, K4 and K7. For HH/HV- and VV/VH-polarized data the Kennaugh Matrix consists of the elements K0, K1, K5 and K6 [31]. Figure 1c shows as a false-color composite of the Kennaugh Matrix elements K1, K2 and K3, which were processed using the quad-polarized R-2 data. For the purpose of comparison Figure 1d shows a Landsat TM true-color RGB composite acquired in summer 2011, concurrent with the R-2 imagery.

Eigen-decomposition—The Eigen-decomposition approach is a frequently used to process PolSAR data [25,26]. It decomposes the incoherent signal (usually stored in the Covariance or Coherency

Matrix) using eigenvalues ($\lambda$) and eigenvectors ($u_x$) ((1) and (2)). In the formula $^H$ denotes the conjugate transpose. Note that the eigenvalues of the Covariance or Coherency Matrix are the same, while the eigenvectors differ. For dual-polarized data two eigenvalues and eigenvectors are obtained (1); while for quad-polarized data three eigenvalues and eigenvectors are obtained (2) when reciprocity is anticipated due to a monostatic acquisition geometry. Consequently, identical scattering from HV and VH is assumed.

$$U = \lambda_1 u_1 u_1^H + \lambda_2 u_2 u_2^H \tag{1}$$

$$U = \lambda_1 u_1 u_1^H + \lambda_2 u_2 u_2^H + \lambda_3 u_3 u_3^H \tag{2}$$

Entropy/Alpha/Anisotropy—The Eigen-decomposition was used to process additional features that describe scattering processes [25,26,42,43]. The polarimetric Entropy and Alpha Scattering angle describes the scattering properties of incoherent (natural) scatterers. Entropy (3) and (4) can be understood as the degree of randomness of the scattered signal and is described by the logarithmic sum of the pseudo probabilities $p$ of the eigenvalues, and ranges from zero to one. The polarimetric Alpha scattering angle is calculated as the sum of the inverse cosine of the absolute value of the first eigenvector element and is weighted by the pseudo probabilities $p$ (5). Cloude and Pottier also showed a third feature for quad-polarized data that is calculated via the ratio between the normalized difference of the second and third eigenvalue: the Anisotropy (6), which indicates the relevance of secondary scattering processes. Anisotropy, understood in the quad-polarimetric sense, is unavailable for dual-polarized data. In the formulas $n$ is equal to two for dual-polarized data and three for quad-polarized data. Note that Entropies of the Covariance or Coherency Matrix are the same, but the Alpha scattering angles are different due to the differences between the Eigenvectors.

$$p_i = \frac{\lambda_i}{\sum_{k=1}^{n} \lambda_k} \tag{3}$$

$$H = -\sum_{i=1}^{n} p_i \log_n(p_i) \tag{4}$$

$$\alpha = \sum_{i=1}^{n} p_i \cos^{-1}(|u_{1i}|) \tag{5}$$

$$A = \frac{\lambda_2 - \lambda_3}{\lambda_2 + \lambda_3} \tag{6}$$

Model-Based Decompositions—Besides the Kennaugh Matrix elements, the polarimetric intensities and the Eigen-decomposition features, three Model-based decompositions were applied to the data which apply simplified, pre-defined scattering models. For the quad-polarized data of R-2 and ALOS-2, the Three Component Yamaguchi Decomposition [27] was applied. This approach decomposes the total backscattered energy $P_{\text{Total}}$ into the intensities of surface scattering ($P_{\text{surface}}$), double bounce scattering ($P_{\text{double bounce}}$) and volume scattering ($P_{\text{volume}}$) (7). This frequently used approach is suitable for comprehending and characterizing predominant scattering processes in nature.

$$P_{\text{Total}} = P_{\text{surface}} + P_{\text{double bounce}} + P_{\text{volume}} \tag{7}$$

As shown by [14] the approach of Yamaguchi can be adopted for HH/VV-polarized data, by decomposing the total backscattered energy $P_{\text{Total}}$ into the intensities of surface scattering ($P_{\text{surface}}$) and double bounce scattering ($P_{\text{double bounce}}$) (8). The correlation between the corresponding features of this decomposition and the Yamaguchi Decomposition are then a function of the presence and power of volume scattering processes [14]. Specifically, features are more highly correlated if volume scattering is negligible.

$$P_{\text{Total}} = P_{\text{surface}} + P_{\text{double bounce}} \tag{8}$$

For HH/VV-polarized data the approach of [30] can be applied as an alternative dual-polarimetric decomposition technique. The approach involves a synthetized HV channel and the polarimetric $H$

(see the preceding paragraph) (see also [44]). This technique decomposes the total backscattered energy $P_{\text{Total}}$ into contributions from scattering from ground ($P_{\text{ground}}$) and from vegetation ($P_{\text{volume}}$) (9).

$$P_{\text{Total}} = P_{\text{volume}} + P_{\text{ground}} \tag{9}$$

The two component decompositions of [14,30] were applied to the X-Band HH/VV data of TSX, to the synthetic HH/VV data of R-2 and ALOS-2. Table 3c lists all the polarimetric data that were used in this study, and provides abbreviations that are used hereafter to refer to each decomposition element. The descriptions of Figures 3 and 4 list all features that were processed for a certain type of polarized data, e.g., for HH/HV or HH/VV data.

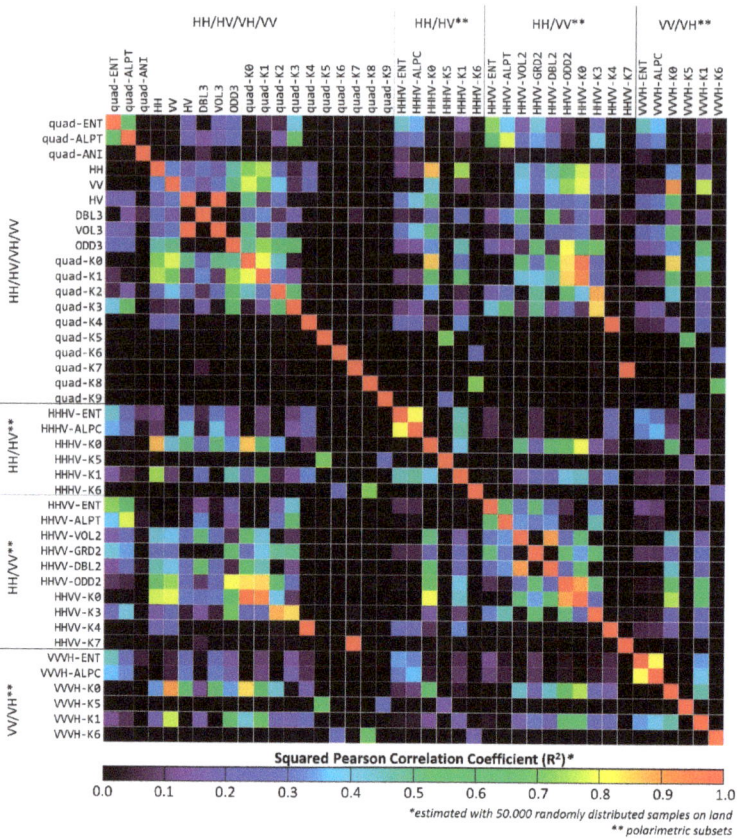

**Figure 3.** Correlation-Matrix of quad- and dual-polarimetric C-Band Radarsat-2 features showing the squared Linear Pearson Correlation Coefficient ($R^2$) ranging from 0.0 (no correlation) to 1.0 (full linear correlation, determination respectively). $R^2$ was estimated using 50,000 randomly distributed samples on land (roughly 10% of all land pixels). Note that dual-polarimetric data of Radarsat-2 were derived as polarimetric subsets and thus are not affected by temporal variations. Feature abbreviations are as follows (see Table 2): ENT (Entropy), ALPT (polarimetric Alpha scattering angle of Coherency Matrix), ALPC (polarimetric Alpha scattering angle of Covariance Matrix), HH/VV/VH (PolSAR Channels), DBL3 (double bounce of the Yamaguchi et al. Decomposition [27]), VOL3 (volume scattering of the Yamaguchi et al. Decomposition), ODD3 (surface scattering of the Yamaguchi et al. Decomposition), K0–K9 (elements of the Kennaugh Matrix [31]), VOL2 (volume scattering of [30]), GRD2 (ground scattering of [30]), DBL2 (double bounce of the [14]), ODD2 (surface scattering of the [14]).

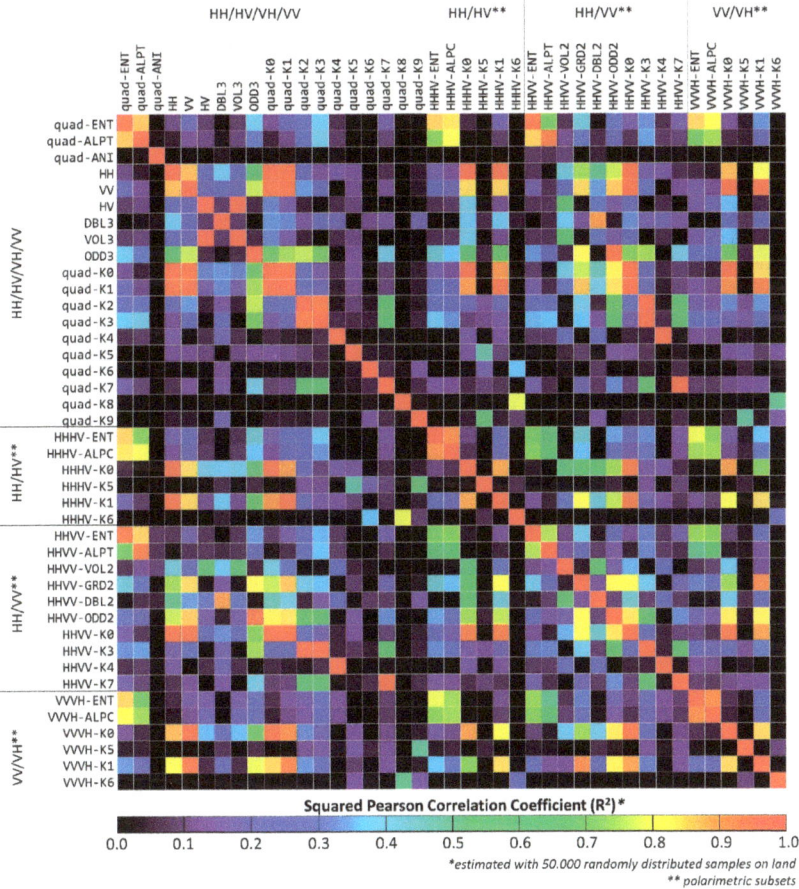

**Figure 4.** Correlation-Matrix of quad- and dual-polarimetric L-Band ALOS-2 features showing the squared Linear Pearson Correlation Coefficient ($R^2$) ranging from 0.0 (no correlation) to 1.0 (full linear correlation, determination respectively). $R^2$ was estimated using 50,000 randomly distributed samples on land (roughly 10% of all land pixels). Note that dual-polarimetric data of ALOS-2 were derived as polarimetric subsets and thus are not affected by temporal variations. Feature abbreviations are as follows (see Table 2): ENT (Entropy), ALPT (polarimetric Alpha scattering angle of Coherency Matrix), ALPC (polarimetric Alpha scattering angle of Covariance Matrix), HH/VV/VH (PolSAR Channels), DBL3 (double bounce of the Yamaguchi et al. Decomposition [27]), VOL3 (volume scattering of the Yamaguchi et al. Decomposition), ODD3 (surface scattering of the Yamaguchi et al. Decomposition), K0–K9 (elements of the Kennaugh Matrix [31]), VOL2 (volume scattering of [30]), GRD2 (ground scattering of [30]), DBL2 (double bounce of [14]), ODD2 (surface scattering of [14]).

## 2.3. Correlation, Class Separability, and Feature Selection

The correlations between the above listed decomposition features were examined using the dB-scaled sigma nought intensity values of the calibrated data, where applicable. Correlations were estimated using a random sample of 50,000 points over land (see Section 2.2.1. Land Cover Reference); thus values and analyses were completed independent of the land cover classes of interest. The squared linear Pearson Correlation Coefficient ($R^2$) was used in all cases. The coefficient $R^2$ is defined as the squared ratio between the covariance (*Cov*) of two variables (*i;j*) and the product of the individual standard deviations $(\sigma_i \sigma_j)$ (10). $R^2$ is frequently used to quantify the degree of determination between two variables, though still can be interpreted as a coefficient that quantifies the correlation. $R^2$ ranges from zero to one; a value of one (zero) indicates perfect (no) linear correlation and a maximum (minimum) determination, 100% (0%) of the explained variance, respectively [45].

$$R^2 = \left( \frac{Cov(i,j)}{\sigma_i \sigma_j} \right)^2 \tag{10}$$

All PolSAR data were then used in separability analyses to quantify the ability of the polarimetric information to discriminate the land cover classes considered in this research. The *Transformed Divergence* (TD) (11) [45,46], *Bhattacharyya Distance* (BD) (12) [47], and *Jefferys Matusita Distance* (JD) (13) [45,46] were processed for each PolSAR feature space and each wavelength for all land cover classes [47]. The features are processed for two classes $c$ and $d$ by assessing the classes' mean vectors $M$ (14) and the classes' covariance matrix $V$ (15) for a given set of features (as a minimum, two features are required). In the formula $tr$ denotes the trace of a matrix, formula $det$ denotes the determinant of a matrix, $^T$ refers to the matrix/vector transpose, and $Cov$ denotes the covariance. The separability features TD and JD have been shown to act as meaningful predictors for classification potential, thus a high separability indicates greater potential for class discrimination [48,49].

$$TD = 2000 \left[ 1 - exp \left( \frac{-0.5 \left( tr \left[ (V_c - V_d)\left( V_d^{-1} - V_c^{-1} \right) \right] + tr \left[ \left( V_c^{-1} + V_d^{-1} \right)(M_c - M_d)(M_c - M_d)^T \right] \right)}{8} \right) \right] \tag{11}$$

$$BD = 0.125(M_c - M_d)^T 0.5(V_c + V_d)(M_c - M_d) + 0.5 log_e \frac{det(0.5(V_c + V_d))}{\sqrt{det(V_c)}\sqrt{det(V_d)}} \tag{12}$$

$$JD = \sqrt{2(1 - e^{-BD})} \tag{13}$$

$$M_c = \begin{pmatrix} \mu_{c1} \\ \mu_{c2} \\ \vdots \\ \mu_{cn} \end{pmatrix} \tag{14}$$

$$V_c = \begin{pmatrix} Cov_c(1,1) & \cdots & Cov_c(1,n) \\ \vdots & \ddots & \vdots \\ Cov_c(n,1) & \cdots & Cov_c(n,n) \end{pmatrix} \tag{15}$$

The metrics TD and JD can further be used for feature selection in order to identify those that are most meaningful for class separation among a given set of features. This can be achieved by calculating the increase in separability (SI) (16): displayed as the amount of separability (*SP*) that is gained when a feature of interest ($x$), e.g., K0, is added to an existing feature space (K). The average increase in separability can be processed by averaging the SI values of each possible feature combination, e.g., the increase in separability when K0 is added to {K1, K2} or {K1, K3} or . . . , {K1, K2, K3} or {K1, K2, K4} or . . . , {K1, K2, K3, K4} or {K1, K2, K3, K5} or . . . , and so on.

$$SI_X = SP_{\{x \cup K\}} - SP_{\{K\}} \tag{16}$$

The separability metrics were employed to demonstrate the differences between the PolSAR features, to gauge their use in classification, and to determine which land cover classes can be separated with the PolSAR features. All of the investigated separability distances require normally distributed data, or at least symmetrically distributed data. Such symmetric distribution properties can be assumed for most of the investigated features.

## 3. Results

### 3.1. Corrleation

Correlations among the decomposition features of dual- and quad-polarized data were investigated prior to the assessment of the backscatter characteristics of the land cover classes and the separability of classes. The squared Pearson Correlation Coefficient ($R^2$) was derived using the 50,000 randomly distributed samples on land, and which represented 10% of all land pixels inside the common coverage of the TSX, ALOS, ALOS-2 and R-2 imagery. Results were drawn as correlation matrices for the features of C-Band data of R-2 in Figure 3, for features of L-Band data of ALOS-2 in Figure 4.

For C-Band data (Figure 3), it was observed that Kennaugh Matrix elements K4, K5, K6, K7, K8 and K9 of the quad-polarized data showed the lowest correlations among all other investigated decomposition features. An explanation for this is that the Kennaugh elements usually are uncorrelated, and that the elements K5 to K8 are phase-based elements, which are more or less stochastically fluctuating over natural targets [31]. The $R^2$ values were less than 0.4 with the exception of K4 and K7. These features were highly correlated with K4 and K7 of HH/VV-polarized data ($R^2$ values of about 0.9). This high degree of correlation is because K4 holds the relation of HH to VV, which is not kept in other decompositions, and the same applies for K7. As well, the correlations between K5 and K5 of HH/HV and VV/VH were moderately high ($R^2$ values of about 0.6). Similarly, the $R^2$ values of K8 and K6 of HH/HV and VV/VH were around 0.5. The reason for these observations are most likely the similar polarimetric behavior (diagonal diattenuation [31]) only with different input channels (HH/HV and VV/VH, respectively).

The same observations were made for the L-Band data of ALOS-2 (Figure 4); however, correlations between the Kennaugh Matrix elements K4, K5, K6, K7, K8 and K9 of quad-, HH/VV-, HH/HV- and VV/VH-polarized data were generally higher. For example, correlation between K7 of quad-polarized Kennaugh Matrix and K3 of HH/VV-polarized Kennaugh Matrix showed $R^2$ values of about 0.5. Among the Kennaugh Matrix elements K0, K1, K2 and K3 of quad-, HH/VV-, HH/HV- and VV/VH-polarized data, the following distinct linear correlations were observed: K0 of quad- and HH/VV-polarized Kennaugh Matrix showed $R^2$ values greater than 0.9 in the C- and L-Band; K3 of quad- and HH/VV-polarized Kennaugh Matrix showed $R^2$ values greater than 0.8 in the C- and L-Band; K0 of quad-, HH/HV- and VV/VH-polarized Kennaugh Matrix showed $R^2$ values greater than 0.8 in the C- and L-Band. Again the correlations between the Kennaugh Matrix elements of quad-, HH/VV-, HH/HV- and VV/VH-polarized data were generally higher in the L- than in the C-Band. Thus, most likely the L-Band data is more "stable" in a polarimetric sense due to a longer wavelength. Further, the high correlation of K0 of quad- and HH/VV-polarized Kennaugh Matrix is present since HH and VV record the vast majority of backscatter, while the HV contribution is negligible.

Among the model-based (power) decomposition features of quad- and HH/VV-polarized data, good correspondence between the DBL3 and DBL2 ($R^2$ values of about 0.8), the ODD3 and ODD2 ($R^2$ values of about 0.7), the VOL3 and VOL2 ($R^2$ values of about 0.7 (L-Band) and 0.4 (C-Band)) and the ODD3 and GRD2 ($R^2$ values of about 0.8 (L-Band) and 0.6 (C-Band)) was observed for both C- and L-Band data. The $R^2$ values between any of the model-based (power) decomposition features and any other polarimetric feature were lower than these observations, with the exception of VOL3 and HV showing $R^2$ values of about 0.95 (C- and L-Band). The reason for this observation can be seen in the low proportion of volume scattering for the tundra environment, making the influence of the cross-polarization component negligible, and decomposition features of quad- and HH/VV-polarized data highly correlated.

With respect to the Eigen-decomposition features, ENT and polarimetric Alpha scattering angles (ALPT/ALPC) were highly correlated between the ENT and the ALPT of HH/VV-polarized data, and this was true for both C- and L-Band ($R^2$ values of about 0.7). Additionally, ENT and ALPT were highly correlated with each other, with $R^2$ values of about 0.7–0.8 (quad- and HH/VV-polarized data) and ENT and ALPC were moderately correlated with $R^2$ values of about 0.4–0.5 (HH/HV- and VV/VH-polarized data).

This is most likely because most reflection is recorded in HH and VV intensities, making their contributions higher than the intensities of HV or VH. This leads to a high correlation between the HH/VV- and quad-polarized decomposition features.

In summary, this assessment indicated that Kennaugh Matrix elements K0, K1, K3, K4 and K7 of quad- and HH/VV-polarized data of C- and L-Band were highly correlated and thus can be used interchangeably in some cases, e.g., for image classification. The Kennaugh Matrix elements K5 and K6 of quad- and cross-polarized data showed lower correlation coefficient values. The correlation was generally higher in the L-band, compared to the C-Band, which is likely a result of less interaction between the incident wave and the vegetation body of the long L-Band microwaves; less volume scattering occurs.

*3.2. Backscatter Characterisics*

Figure 5 provides boxplots—showing the minimum, lower quartile (25%), median (50%), upper quartile (75%), maximum—of the land cover classes (see Table 2) for select polarimetric features of X-Band (TSX), C-Band (R-2) and L-Band (ALOS and ALOS-2). Figure 5a–i display the backscatter characteristics of the land cover classes concerning the HH, HV and VV sigma nought intensities in decibels (dB). Figure 5j–o shows the information of the model-based (power) decomposition features of the Yamaguchi Decomposition of C-Band (R-2) and L-Band (ALOS-2) as DBLB3, VOL3 and ODD3; in dB. Figure 5p–x shows the boxplots of the land cover classes for the Kennaugh Matrix elements K0, K3 and K4 of HH/VV-polarized X-Band (TSX), quad-polarized C-Band (R-2) and quad-polarized L-Band (ALOS-2) data in dB.

The class OL showed a unique range of intensity values in the K0 and VV of X- and C- Band, the VOL3 and ODD3 of C-Band, and the K4 and HV of X-, C- and L-Band data. The scattering differences between water and land were clearly pronounced, as water was generally characterized by a low intensity value. This is because water surface was relatively calm, thus it was not observed as rough; the shallow angle incidence angles of the X- and C-Band data, and the longer wavelength of the L-Band. If the water surface were to become rough due to higher wind speeds, higher intensity values of K0 would be observed, thus complicating the separation of the classes. In such cases it is assumed that K3 and K4 will still be suitable to separate land from water, since both are indicators for double bounce scattering, typically minimal for water. Further, BS and BM showed increased HH, VV, HV, DBL3 and ODD3 scattering at C- and L-Band compared to other classes. The range of values observed for the BW class was more unique for L-Band features, than X- and C-Band features. The most distinct values were for L-Band VOL3; as the BW's median value exceeded +5 dB, compared to the median value of all other land cover classes. Scattering from BW at L-Band is therefore characterized by high intensity values for HH, VOL3 and ODD3, with medians of about −10 dB and −7 dB; whereas at C-Band the scattering from BW is characterized by high HH and ODD3 intensities. The median intensity of BW is comparably low in X-Band HH and differences of BW's statistics to BS and HT are less pronounced. Thus, independent of wavelength and decomposition technique, the data were sensitive to scattering differences between land and water (OL), and between sandy bare ground (BS) and mixed non-vegetated sediment (BM). BM had higher backscatter than BS, which can be attributed to the higher surface roughness of BM (grain sizes of 2.0 mm–256.0 mm) compared to BS (grain sizes of 0.0625 mm–2.0 mm), which leads to higher backscatter intensities.The difference between HT and ST scattering is characterized by increased HV, VOL3, and K0 intensities at X- and C-Band; however, there is substantial overlap in their distributions and the differences between median values were small; ranging from +2 dB (X-Band HV) to +3 dB (C-Band VOL3) between HT and ST. The largest differences in HT and ST statistics were nevertheless found for VOL3 of the C-Band data, but the data ranges of HT and ST (lower quantile to upper quantile) also overlap the ranges of BM, BW, WI and WT. Therefore, the X- and C-Band showed higher volume scattering intensities from shrub dominated tundra (ST) compared to herb dominated tundra (HT). This is likely due to a higher proportion of volume scattering in the shrub plants, which is caused by the relatively short wavelength. Contrarily, the L-Band HV and volume scattering intensities (VOL3) were not sensitive to this difference.

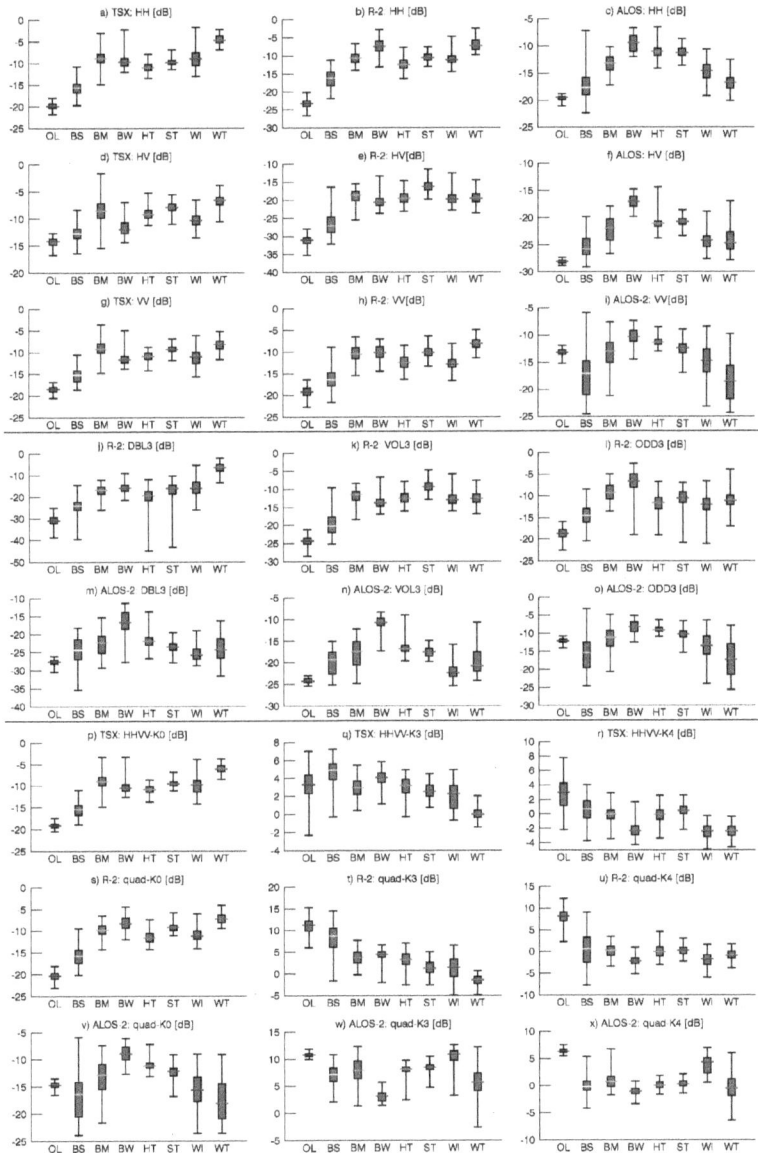

**Figure 5.** Boxplots (minimum, lower quartile (25%), median (50%), upper quartile (75%), maximum) of (**a–i**) the polarimetric channels (HH/HV/VV) of TSX, R-2, ALOS and ALOS-2; (**j–o**) Yamaguchi Decomposition features (DBL3/VOL3/ODD3) of R-2 and ALOS-2; (**p–x**) Kennaugh Matrix elements (K0/K3/K4) of TSX, R-2 and ALOS-2 for the land cover classes OL (Permanent Water Bodies), BS (Sand), BM (Mixed Sediment), BW (Driftwood Accumulation), HT (Herb Dominated Tundra), ST (Shrub Dominated Tundra), WI (Inundated Low Lying Tundra) and WT (Wetland) (see Table 2).

The L-Band HV channel and the volume scattering intensity of the Yamaguchi decomposition showed the same median values and comparable data ranges (lower to upper quantile) for these two land cover classes. At L-Band, the difference between HT and ST was better expressed via the VV channel, and the surface scattering intensity of the Yamaguchi decomposition via lower intensities of the ST compared to HT. This indicates an absence of volume scattering processes and a full penetration of the vegetation by the L-Band microwaves. Assuming that this observation is not caused by temporal variations present in the ALOS-2 data—which were acquired in 2016, while C-, X-Band and the land cover reference data were acquired in 2010/2011, the signal can be interpreted to represent backscattering mostly from the ground, as it is assumed that major changes in land cover type present have not occurred in this time.

The differences between the land cover classes WI and WT were characterized by increased DBL3, K0, HH and VV scattering in X- and C-Band and by K3 in X-, C- and L-Band. The statistics of WT showed a clear separation from the other land cover classes in the DBL3 and K3 of C-Band and the HH and K3 of X-Band. The difference between the WT's median value and the median value of any other land cover class exceeded +5 dB in the X-Band HH and C-Band DBL3. The differences between wetland (WT) and inundated low-lying tundra (WI) was observed as higher HH and VV intensities—and the Kennaugh Matrix element K4 accordingly—in X- and C-Band and the double bounce intensity of the Yamaguchi decomposition. Further, both classes were characterized by comparably low values of the Kennaugh Matrix element K3, which points to distinct double bounce scattering, since $\Re\left(S_{HH}S_{VV}^{*}\right)$ is a known discriminator for this type of scattering (compare [27,31]). In contrast, both classes showed low intensities in the HH, VV and double bounce of the Yamaguchi decomposition at L-Band. Accordingly, K3 and K4 were less distinct and no double bounce scattering was present, when using the L-Band, which again is most likely due to the relatively short statured vegetation, the high penetration depth, and the absence of interactions between incident microwaves and the water surface and vegetation canopy.

In addition to the boxplots, Figure 6 shows scatterplots of the Kennaugh Matrix elements K0, K3 and K4 of X-, C- and L-Band data in order to investigate the scattering characteristics of the land cover classes in a multivariate feature space. Figure 6a–i shows the position of the land cover reference in the K0/K3 (left column), K0/K4 (center column) and K3/K4 (right column) feature spaces of X-Band (a–c), C-Band (d–f) and L-Band (g–i). These results show that the feature space K0/K3 of X-and C-Band facilitates the differentiation of the classes: OL, BS and WT; however, the position of values for BM, HT, ST and WI were indiscriminant from others. The feature space K0/K4 (Figure 6b,e,h) shows increased distance between samples of HT/ST and WI—especially at X-Band; however, a substantial degree of overlap between the samples of BM and BW, and the samples of HT, ST and WI was present. K0 provided the best separation between land cover classes at X-Band, C-Band, and L-Band. This is unsurprising since K0 of X- and C-Band showed a high positive linear correlation ($R^2$ of 0.8) (Figure 6j). The combination of short- and longwave SAR facilitated the separation of the WI, BM and BW samples, and the features of X-/C- and L-Band showed no linear correlation ($R^2$ less than 0.1) (Figure 6k,l).

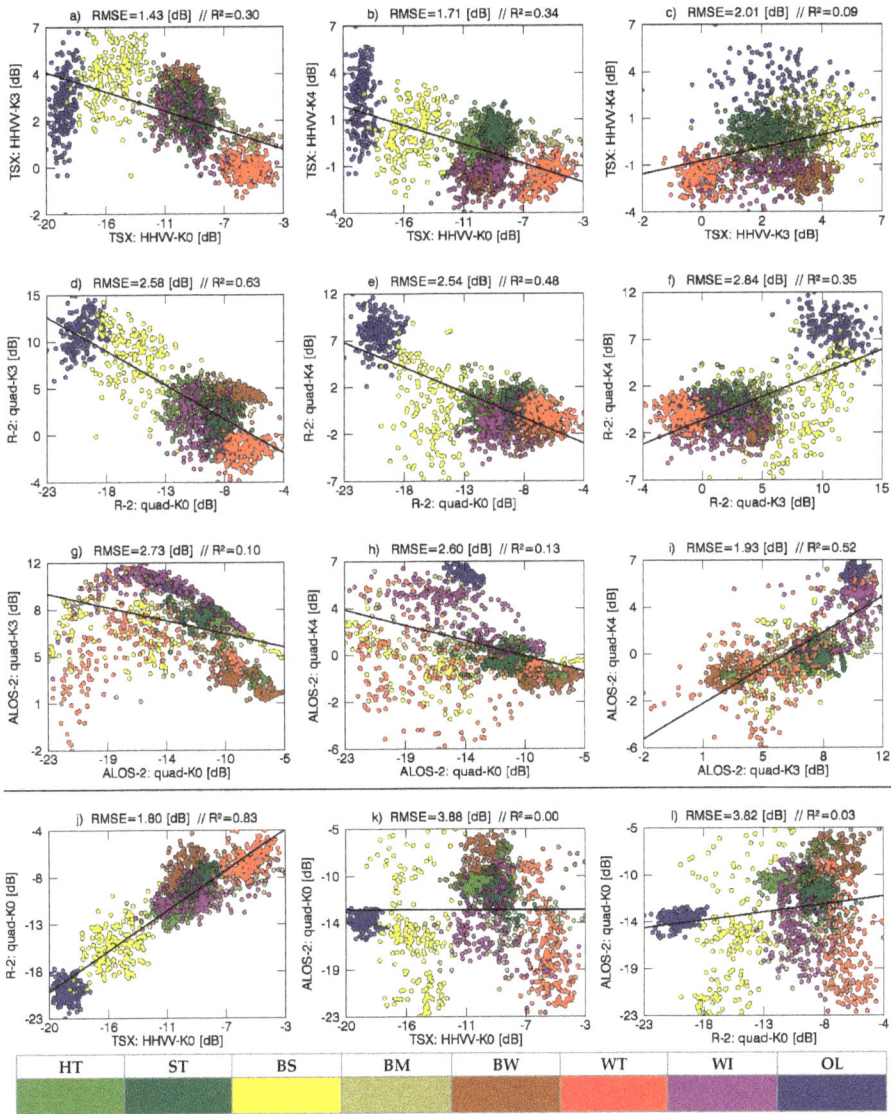

**Figure 6.** Scatterplots of the land cover classes OL (Permanent Water Bodies), BS (Sand), BM (Mixed Sediment), BW (Driftwood Accumulation), HT (Herb Dominated Tundra), ST (Shrub Dominated Tundra), WI (Inundated Low Lying Tundra) and WT (Wetland) (see Table 2) for: (**a–c**) Kennaugh elements K0/K3/K4 of HHVV-polarized X-Band (TSX); (**d–f**) Kennaugh elements K0/K3/K4 of quad-polarized C-Band (R-2); (**g–i**) Kennaugh elements K0/K3/K4 of quad-polarized L-Band (ALOS-2); (**j–l**) Kennaugh element K0 of TSX, R-2 and ALOS-2. In the figure title the Root-Means-Square-Error (RMSE) and the squared Linear Pearson Correlation Coefficient ($R^2$) are drawn.

In summary, the X-, C- and L-Band data exhibit distinct scattering characteristics for the different land cover classes. All PolSAR data were sensitive to the OL, BS and BM coverage; additionally, L-Band data were most sensitive to the BW. The X- and C-Band features were suited to pronounce differences in WI and WT, and HT and ST coverage via the features HV, VOL3, DBL3, K0 and K3.

### 3.3. Class Separability and Feature Selection

Multivariate assessment was completed for all feature spaces of the decomposition elements of dual- and quad-polarized X-, C- and L-Band datasets. The separabilities between classes (for the feature spaces of interest) were quantified for all possible combinations of variables (41) using the Transformed Divergence (TD) (Table 4) and the squared Jefferys Matusita Distance (JD) (Table 5). The feature spaces were ranked in descending order based on the average separability (AV) by the feature space of interest. For the purpose of comparison the tables also list separabilities achieved with the multispectral Landsat TM data using the six spectral bands (thermal information was excluded).

Results showed that the quad-polarized Kennaugh Matrices of ALOS-2 and R-2 offered the best separation of all land cover classes, followed by the HH/VV-polarized Kennaugh Matrices of ALOS-2, R-2 and TSX. The separability distances TD and JD further indicated that AV of ALOS-2 was comparable to the AV offered by multispectral data. As well, among the different PolSAR decompositions, the use of all Kennaugh Matrix elements was more beneficial for class separation than using the features of the model-based (power) Decompositions, Eigen-decompositions, or the intensities of the polarimetric channels. For C- and L-Band it was further observed that TD and JD of the Kennaugh Matrix decreased from quad-, to HH/VV-, to VV/VH- to HH/HV-polarized data.

For X-Band it was observed that TD and JD of the Kennaugh Matrix decreased from HH/VV- to HH/HV-polarized data. The separability of the Eigen-decomposition features ENT, ALPT or ALPC and ANI was low, and these feature spaces, as indicated by JD, offered the lowest separability between classes among all investigated feature spaces; independent of the wavelength (X-, C- or L-Band). With HH/VV-polarized data, the high correlation of Entropy and the Alpha scattering angles was observed by others [50], though. Another reason for this might be the lack of diversity of scattering processes in this rather "bare" landscape, thus the Entropy/Alpha feature space remains "unfilled" to a certain degree. Specifically, the tundra landscape examined in this research, offers a minor depolarizing, and low entropy environment.

The separability distance JD further outlined that the average class separability decreased from ENT/ALPT/ANI (quad) to ENT/ALPT (HH/VV) to ENT/ALPC (VV/VH) to ENT/ALPC (HH/HV). This might simply be a function of intensity, which decreased from quad to HH/VV to VV/VH to HH/HV, since with lower intensities there is also lower information content. Among the land cover classes the classes OL, BS, BW and WT were shown to be the land cover classes with the highest average separability, thus the PolSAR data were especially suited to characterize these classes. The lowest average separability was observed for the land cover classes BM and WI, while separability of HT and ST was moderately high.

Table 6 draws the average increase of the separability features TD and JD for the ten elements of the quad-polarized Kennaugh Matrices (K0–K9) of C-Band (left column) and L-Band (right column) data. This assessment indicates that the average increase in separability when a feature of interest (K0–K9) is added to an existing feature space. This metric was used to identify the most important elements of the Kennaugh Matrix for class separation. For both C- and L-Band it was observed that K0, K1, K2, K3 and K4 were more important for class separation than K5, K6, K7, K8 and K9. An explanation for this observation is that K0, K1 and K2 are intensity-based elements and are thus value are generally stable. K3 and K4 also use the cross-polarized intensity and the co-polarized phase, and are therefore relatively stable as well. Contrarily, K5, K6, K7, K8 and K9 are phase-based elements and therefore generally unstable in natural environments. With respect to the Coherency Matrix, K0 to K4 explain the diagonal elements which can be associated with dominant scattering processes, including: surface, double bounce and volume scattering. Hence (K5–K9, representing the off-diagonal elements of the Coherency Matrix, have minor relevance for the class separability.

**Table 4.** Average class separability measured as Transformed Divergence (TD) for classes OL (Permanent Water Bodies), BS (Sand), BM (Mixed Sediment), BW (Driftwood Accumulation), HT (Herb Dominated Tundra), ST (Shrub Dominated Tundra), WI (Inundated Low Lying Tundra) and WT (Wetland) (see Table 2). TD is ranging from 0 to 2000; higher values indicate better class separation. The feature spaces are ranked in descending order of the average separability (AV). The black bars are scaled linearly between the minimum and maximum of AV. The colors from red to yellow to green correspond to the 10%, 50% and 90% quantiles of the AV.

| # | SENSOR | POL. | FEATURES | OL | BS | BM | BW | HT | ST | WI | WT | AV |
|---|--------|------|----------|----|----|----|----|----|----|----|----|----|
| 1 | ALOS-2 | quad | K-Matrix | 2000 | 1990 | 1963 | 1999 | 1999 | 2000 | 1973 | 2000 | 1991 |
| 2 | Landsat TM | n/a | (Band 1-5 & 7) | 2000 | 1925 | 1925 | 2000 | 1993 | 1921 | 1926 | 2000 | 1961 |
| 3 | ALOS-2 | HHVV | K-Matrix | 2000 | 1763 | 1593 | 1952 | 1930 | 1868 | 1849 | 1881 | 1854 |
| 4 | R-2 | quad | K-Matrix | 2000 | 1993 | 1640 | 1908 | 1680 | 1731 | 1782 | 1999 | 1842 |
| 5 | ALOS-2 | quad | HH/HV/VV | 2000 | 1550 | 1439 | 1929 | 1796 | 1766 | 1817 | 1636 | 1742 |
| 6 | ALOS | HHHV | K-Matrix | 2000 | 1681 | 1516 | 1914 | 1645 | 1717 | 1652 | 1740 | 1733 |
| 7 | ALOS-2 | VVVH | K-Matrix | 2000 | 1521 | 1394 | 1961 | 1743 | 1621 | 1736 | 1661 | 1705 |
| 8 | TSX | HHVV | K-Matrix | 1998 | 1934 | 1447 | 1574 | 1464 | 1605 | 1620 | 1966 | 1701 |
| 9 | ALOS-2 | quad | DBL3/VOL3/ODD3 | 1999 | 1433 | 1353 | 1938 | 1787 | 1685 | 1696 | 1558 | 1681 |
| 10 | ALOS-2 | HHHV | K-Matrix | 2000 | 1381 | 1334 | 1951 | 1838 | 1675 | 1514 | 1556 | 1656 |
| 11 | ALOS-2 | VVVH | VV/VH | 1999 | 1384 | 1265 | 1929 | 1680 | 1582 | 1657 | 1520 | 1627 |
| 12 | R-2 | HHVV | K-Matrix | 1998 | 1805 | 1282 | 1671 | 1306 | 1395 | 1442 | 1982 | 1610 |
| 13 | TSX | HHHV | K-Matrix | 2000 | 1677 | 1320 | 1566 | 1488 | 1554 | 1311 | 1891 | 1601 |
| 14 | ALOS-2 | HHHV | HH/HV | 1995 | 1318 | 1211 | 1898 | 1727 | 1628 | 1538 | 1428 | 1593 |
| 15 | R-2 | quad | DBL3/VOL3/ODD3 | 1992 | 1875 | 1206 | 1529 | 1320 | 1633 | 1288 | 1893 | 1592 |
| 16 | ALOS-2 | HHVV | DBL2/ODD2 | 1997 | 1315 | 1177 | 1871 | 1745 | 1580 | 1465 | 1486 | 1579 |
| 17 | ALOS-2 | HHVV | VOL2/GRD2 | 1995 | 1292 | 1188 | 1829 | 1770 | 1615 | 1433 | 1491 | 1577 |
| 18 | ALOS | HHHV | HH/HV | 2000 | 1464 | 1249 | 1848 | 1410 | 1609 | 1349 | 1605 | 1567 |
| 19 | TSX | HHVV | HH/VV | 1995 | 1883 | 1302 | 1289 | 1340 | 1530 | 1340 | 1848 | 1566 |
| 20 | TSX | HHVV | HH/VV | 1998 | 1657 | 1268 | 1401 | 1418 | 1485 | 1261 | 1844 | 1542 |
| 21 | ALOS-2 | HHVV | HH/VV | 2000 | 1401 | 1073 | 1562 | 1469 | 1462 | 1582 | 1476 | 1503 |
| 22 | R-2 | quad | HH/HV/VV | 1996 | 1865 | 1134 | 1459 | 1317 | 1464 | 1219 | 1561 | 1502 |
| 23 | R-2 | HHHV | K-Matrix | 1997 | 1849 | 1057 | 1362 | 1240 | 1429 | 1062 | 1560 | 1445 |
| 24 | TSX | HHVV | DBL2/ODD2 | 1995 | 1834 | 1046 | 1151 | 1187 | 1327 | 1030 | 1747 | 1415 |
| 25 | R-2 | HHVV | DBL2/ODD2 | 1991 | 1755 | 984 | 1332 | 1136 | 1139 | 1031 | 1742 | 1389 |
| 26 | TSX | HHVV | ENT/ALPC | 1980 | 1344 | 1164 | 1218 | 1422 | 1308 | 1279 | 1390 | 1388 |
| 27 | R-2 | HHHV | HH/HV | 1994 | 1844 | 970 | 1322 | 1221 | 1378 | 969 | 1334 | 1379 |
| 28 | TSX | HHVV | VOL2/GRD2 | 1990 | 1811 | 1009 | 1009 | 1082 | 1235 | 983 | 1899 | 1377 |
| 29 | ALOS-2 | quad | ENT/ALPT/ANI | 2000 | 1097 | 1001 | 1541 | 1339 | 1486 | 1245 | 1306 | 1377 |
| 30 | R-2 | VVVH | K-Matrix | 1990 | 1790 | 930 | 1060 | 1083 | 1327 | 1115 | 1535 | 1354 |
| 31 | R-2 | VVVH | VV/VH | 1985 | 1803 | 882 | 1077 | 1100 | 1321 | 1076 | 1377 | 1328 |
| 32 | R-2 | HHVV | HH/VV | 1989 | 1672 | 970 | 1230 | 1087 | 1143 | 1051 | 1439 | 1322 |
| 33 | R-2 | HHVV | VOL2/GRD2 | 1988 | 1729 | 909 | 1148 | 1075 | 1023 | 948 | 1573 | 1299 |
| 34 | ALOS-2 | HHVV | ENT/ALPT | 2000 | 997 | 862 | 1426 | 1255 | 1281 | 1036 | 1264 | 1265 |
| 35 | ALOS-2 | VVVH | ENT/ALPC | 2000 | 923 | 841 | 1763 | 868 | 979 | 1312 | 1152 | 1230 |
| 36 | R-2 | quad | ENT/ALPT/ANI | 1436 | 959 | 896 | 1110 | 1043 | 1241 | 1119 | 1924 | 1216 |
| 37 | R-2 | HHVV | ENT/ALPT | 1433 | 794 | 800 | 852 | 810 | 1009 | 1060 | 1911 | 1084 |
| 38 | TSX | HHVV | ENT/ALPT | 1016 | 764 | 738 | 1234 | 807 | 830 | 1385 | 1788 | 1070 |
| 39 | ALOS | HHHV | ENT/ALPC | 1445 | 1016 | 722 | 1123 | 877 | 840 | 875 | 1546 | 1056 |
| 40 | ALOS-2 | HHHV | ENT/ALPC | 1580 | 814 | 599 | 1232 | 920 | 751 | 716 | 1019 | 954 |
| 41 | R-2 | VVVH | ENT/ALPC | 1291 | 721 | 567 | 681 | 915 | 1131 | 849 | 914 | 883 |
| 42 | R-2 | HHHV | ENT/ALPC | 571 | 593 | 489 | 1058 | 838 | 1133 | 493 | 1092 | 784 |

**Table 5.** Average class separability measured as squared Jefferys Matusita Distance (JD) for classes OL (Permanent Water Bodies), BS (Sand), BM (Mixed Sediment), BW (Driftwood Accumulation), HT (Herb Dominated Tundra), ST (Shrub Dominated Tundra), WI (Inundated Low Lying Tundra) and WT (Wetland) (see Table 2). JD is ranging from 0 to 2; higher values indicate better class separation. The feature spaces are ranked in descending order of the average separability (AV). The black bars are scaled linearly between the minimum and maximum of AV. The colors from red to yellow to green correspond to the 10%, 50% and 90% quantiles of the AV.

| # | SENSOR | POL. | FEATURES | OL | BS | BM | BW | HT | ST | WI | WT | AV |
|---|--------|------|----------|----|----|----|----|----|----|----|----|----|
| 1 | ALOS-2 | quad | K-Matrix | 2.00 | 1.96 | 1.90 | 1.98 | 1.98 | 1.98 | 1.95 | 2.00 | 1.97 |
| 2 | Landsat TM | n/a | (Band 1-5 & 7) | 2.00 | 1.91 | 1.84 | 2.00 | 1.94 | 1.83 | 1.83 | 2.00 | 1.92 |
| 3 | R-2 | quad | K-Matrix | 2.00 | 1.97 | 1.57 | 1.86 | 1.61 | 1.67 | 1.69 | 1.99 | 1.80 |
| 4 | ALOS-2 | HHVV | K-Matrix | 1.99 | 1.51 | 1.31 | 1.82 | 1.69 | 1.60 | 1.64 | 1.64 | 1.65 |
| 5 | TSX | HHVV | K-Matrix | 1.97 | 1.86 | 1.33 | 1.51 | 1.39 | 1.52 | 1.46 | 1.87 | 1.61 |
| 6 | ALOS | HHHV | K-Matrix | 1.90 | 1.32 | 1.31 | 1.82 | 1.43 | 1.50 | 1.40 | 1.60 | 1.53 |
| 7 | R-2 | HHVV | K-Matrix | 1.97 | 1.68 | 1.22 | 1.58 | 1.24 | 1.32 | 1.31 | 1.95 | 1.53 |
| 8 | ALOS-2 | quad | HH/HV/VV | 1.99 | 1.33 | 1.13 | 1.80 | 1.44 | 1.45 | 1.66 | 1.43 | 1.53 |
| 9 | TSX | HHVV | HH/VV | 1.96 | 1.79 | 1.20 | 1.25 | 1.26 | 1.45 | 1.20 | 1.76 | 1.48 |
| 10 | ALOS-2 | quad | DBL3/VOL3/ODD3 | 1.95 | 1.21 | 1.06 | 1.81 | 1.40 | 1.42 | 1.54 | 1.38 | 1.47 |
| 11 | ALOS-2 | VVVH | K-Matrix | 1.92 | 1.23 | 1.08 | 1.82 | 1.40 | 1.34 | 1.43 | 1.44 | 1.46 |
| 12 | TSX | HHHV | K-Matrix | 1.96 | 1.55 | 1.07 | 1.39 | 1.27 | 1.41 | 1.16 | 1.80 | 1.45 |
| 13 | R-2 | quad | DBL3/VOL3/ODD3 | 1.94 | 1.68 | 1.09 | 1.44 | 1.16 | 1.45 | 1.17 | 1.67 | 1.45 |
| 14 | ALOS-2 | HHHV | K-Matrix | 1.89 | 1.17 | 1.06 | 1.84 | 1.49 | 1.37 | 1.29 | 1.40 | 1.44 |
| 15 | R-2 | quad | HH/HV/VV | 1.97 | 1.70 | 1.07 | 1.35 | 1.20 | 1.37 | 1.15 | 1.49 | 1.41 |
| 16 | TSX | HHHV | HH/HV | 1.95 | 1.52 | 1.00 | 1.29 | 1.23 | 1.38 | 1.13 | 1.76 | 1.41 |
| 17 | ALOS | HHHV | HH/HV | 1.85 | 1.14 | 1.08 | 1.73 | 1.23 | 1.40 | 1.18 | 1.48 | 1.39 |
| 18 | ALOS-2 | VVVH | VV/VH | 1.96 | 1.10 | 0.96 | 1.75 | 1.27 | 1.28 | 1.47 | 1.29 | 1.39 |
| 19 | R-2 | HHHV | K-Matrix | 1.97 | 1.65 | 0.99 | 1.30 | 1.15 | 1.33 | 0.98 | 1.46 | 1.35 |
| 20 | ALOS-2 | HHVV | DBL2/ODD2 | 1.93 | 1.05 | 0.92 | 1.69 | 1.29 | 1.25 | 1.23 | 1.29 | 1.33 |
| 21 | R-2 | HHVV | DBL2/ODD2 | 1.93 | 1.59 | 0.96 | 1.28 | 1.09 | 1.07 | 0.99 | 1.71 | 1.33 |
| 22 | TSX | HHVV | DBL2/ODD2 | 1.94 | 1.74 | 0.93 | 1.12 | 1.12 | 1.21 | 0.88 | 1.63 | 1.32 |
| 23 | ALOS-2 | HHHV | HH/HV | 1.86 | 1.05 | 0.89 | 1.76 | 1.30 | 1.26 | 1.28 | 1.17 | 1.32 |
| 24 | ALOS-2 | HHVV | VOL2/GRD2 | 1.87 | 1.02 | 0.89 | 1.67 | 1.31 | 1.25 | 1.16 | 1.23 | 1.30 |
| 25 | R-2 | HHHV | HH/HV | 1.96 | 1.64 | 0.91 | 1.21 | 1.11 | 1.28 | 0.91 | 1.27 | 1.29 |
| 26 | TSX | HHVV | VOL2/GRD2 | 1.93 | 1.71 | 0.88 | 0.95 | 1.03 | 1.13 | 0.84 | 1.75 | 1.28 |
| 27 | R-2 | HHVV | HH/VV | 1.96 | 1.56 | 0.94 | 1.20 | 1.05 | 1.07 | 1.03 | 1.39 | 1.28 |
| 28 | ALOS-2 | HHVV | HH/VV | 1.96 | 1.09 | 0.87 | 1.32 | 1.15 | 1.17 | 1.40 | 1.17 | 1.27 |
| 29 | R-2 | VVVH | K-Matrix | 1.91 | 1.56 | 0.88 | 1.01 | 1.02 | 1.25 | 1.04 | 1.44 | 1.26 |
| 30 | ALOS-2 | quad | ENT/ALPT/ANI | 2.00 | 0.93 | 0.89 | 1.40 | 1.20 | 1.29 | 1.13 | 1.16 | 1.25 |
| 31 | R-2 | VVVH | VV/VH | 1.90 | 1.54 | 0.84 | 1.00 | 1.01 | 1.26 | 1.01 | 1.32 | 1.23 |
| 32 | R-2 | HHVV | VOL2/GRD2 | 1.93 | 1.57 | 0.86 | 1.09 | 1.02 | 0.96 | 0.88 | 1.48 | 1.22 |
| 33 | ALOS-2 | HHVV | ENT/ALPT | 1.99 | 0.83 | 0.77 | 1.27 | 1.10 | 1.09 | 0.95 | 1.14 | 1.14 |
| 34 | TSX | HHHV | ENT/ALPC | 1.85 | 0.94 | 0.83 | 0.95 | 1.02 | 0.96 | 0.94 | 1.16 | 1.08 |
| 35 | ALOS-2 | VVVH | ENT/ALPC | 1.92 | 0.76 | 0.73 | 1.47 | 0.76 | 0.86 | 1.14 | 0.99 | 1.08 |
| 36 | R-2 | quad | ENT/ALPT/ANI | 1.33 | 0.84 | 0.79 | 0.95 | 0.89 | 1.04 | 0.93 | 1.81 | 1.07 |
| 37 | R-2 | HHVV | ENT/ALPT | 1.36 | 0.73 | 0.73 | 0.77 | 0.74 | 0.88 | 0.88 | 1.80 | 0.99 |
| 38 | TSX | HHVV | ENT/ALPT | 0.83 | 0.60 | 0.63 | 1.02 | 0.61 | 0.62 | 0.84 | 1.57 | 0.84 |
| 39 | ALOS | HHHV | ENT/ALPC | 0.89 | 0.65 | 0.51 | 0.81 | 0.66 | 0.58 | 0.65 | 1.07 | 0.73 |
| 40 | ALOS-2 | HHHV | ENT/ALPC | 1.14 | 0.57 | 0.46 | 0.98 | 0.64 | 0.55 | 0.52 | 0.81 | 0.71 |
| 41 | R-2 | VVVH | ENT/ALPC | 1.05 | 0.48 | 0.45 | 0.51 | 0.73 | 0.89 | 0.69 | 0.75 | 0.69 |
| 42 | R-2 | HHHV | ENT/ALPC | 0.47 | 0.47 | 0.41 | 0.86 | 0.66 | 0.92 | 0.41 | 0.91 | 0.64 |

For C-Band it was further observed that K0 and K1 offered the highest increase for the separation of ST, while for L-Band K1 and K3 were more important for the separation of this land cover class. At C-Band, the information of K2 and K3 was beneficial for the separation of BW, WI and WT. The land cover class HT was best characterized by the elements K0 and K2 at C-Band and K0, K1 and K3 at L-Band. This means that K0 (total backscattered intensity), K1 (absorption element showing the difference between co- and cross-polarized intensities), K2 and K3 (absorption elements that describe the loss of polarization during the scattering process) and K4 (diattenuation element showing the difference between HH and VV intensities) are good descriptors for the examined tundra land cover classes. The elements K7, K8 and K9 (descriptors of the phase delay during the scattering in a certain direction) play a minor role in the separation of classes, as phase delays happen during volume propagation. Since tundra vegetation has a relatively short stature (height), phase delays due to volume propagation are less likely.

**Table 6.** Average increase of the separability features: (a–b) Transformed Divergence (TD) and (c–d) squared Jefferys Matusita Distance (JD) for Kennaugh-Matrix elements K0-K9 of quad-polarimetric C-Band Radarsat-2 (left column) and L-Band ALOS-2 (right column) data and for classes OL (Permanent Water Bodies), BS (Sand), BM (Mixed Sediment), BW (Driftwood Accumulation), HT (Herb Dominated Tundra), ST (Shrub Dominated Tundra), WI (Inundated Low Lying Tundra) and WT (Wetland) (see Table 2). The features are ranked in descending order of the average increase in separability (AV). The last column displays the AV in percent (%) The colors from red to yellow to green correspond to the 10%, 50% and 90% quantiles of the AV data range. The metric displays the average increase in separability when a feature of interest (K0–K9) is added to an existing feature space (see Section 2.3 *Correlation, Class Separability and Feature Selection*).

| C-Band Radarsat-2 | | | | | | | | | | | L-Band ALOS-2 | | | | | | | | | | |
|---|---|---|---|---|---|---|---|---|---|---|---|---|---|---|---|---|---|---|---|---|---|
| (a) Transformed Divergence (TD) | | | | | | | | | | | (b) Transformed Divergence (TD) | | | | | | | | | | |
| | OL | BS | BM | BW | HT | ST | WI | WT | AV | % | | OL | BS | BM | BW | HT | ST | WI | WT | AV | % |
| K0 | 100 | 193 | 150 | 183 | 226 | 168 | 186 | 82 | **161** | 49.8 | K0 | 30 | 115 | 172 | 65 | 109 | 91 | 92 | 56 | **91** | 10.2 |
| K2 | 99 | 129 | 137 | 143 | 182 | 129 | 224 | 163 | **151** | 48.8 | K1 | 32 | 84 | 177 | 84 | 86 | 115 | 99 | 43 | **90** | 10.1 |
| K3 | 90 | 107 | 133 | 170 | 111 | 128 | 152 | 123 | **127** | 39.2 | K3 | 34 | 86 | 162 | 83 | 89 | 100 | 103 | 43 | **87** | 9.9 |
| K1 | 100 | 207 | 101 | 90 | 117 | 173 | 162 | 37 | **123** | 35.6 | K2 | 28 | 74 | 143 | 68 | 83 | 86 | 78 | 34 | **74** | 8.4 |
| K4 | 97 | 92 | 63 | 116 | 61 | 73 | 118 | 19 | **80** | 28.2 | K4 | 36 | 47 | 103 | 35 | 42 | 84 | 145 | 30 | **65** | 7.7 |
| K9 | 2 | 10 | 63 | 92 | 34 | 37 | 45 | 23 | **38** | 5.0 | K7 | 23 | 74 | 97 | 23 | 83 | 53 | 76 | 53 | **60** | 6.8 |
| K7 | 1 | 14 | 29 | 32 | 30 | 22 | 45 | 94 | **33** | 4.9 | K9 | 14 | 86 | 65 | 36 | 25 | 27 | 24 | 38 | **40** | 4.0 |
| K5 | 5 | 11 | 52 | 49 | 29 | 22 | 37 | 36 | **30** | 4.8 | K6 | 13 | 34 | 41 | 9 | 29 | 24 | 21 | 30 | **25** | 2.7 |
| K8 | 6 | 12 | 19 | 49 | 32 | 23 | 21 | 36 | **25** | 4.4 | K5 | 16 | 21 | 34 | 21 | 18 | 23 | 24 | 13 | **21** | 2.2 |
| K6 | 2 | 7 | 16 | 46 | 27 | 13 | 14 | 11 | **17** | 2.3 | K8 | 10 | 22 | 38 | 10 | 11 | 15 | 18 | 32 | **20** | 2.0 |
| (c) Jefferys Matusita Distance (JD) | | | | | | | | | | | (d) Jefferys Matusita Distance (JD) | | | | | | | | | | |
| | OL | BS | BM | BW | HT | ST | WI | WT | AV | % | | OL | BS | BM | BW | HT | ST | WI | WT | AV | % |
| K0 | 0.11 | 0.29 | 0.15 | 0.19 | 0.22 | 0.18 | 0.19 | 0.11 | **0.18** | 57.2 | K3 | 0.09 | 0.16 | 0.24 | 0.14 | 0.16 | 0.16 | 0.20 | 0.13 | **0.16** | 18.3 |
| K2 | 0.10 | 0.19 | 0.13 | 0.13 | 0.16 | 0.13 | 0.20 | 0.21 | **0.16** | 51.9 | K0 | 0.07 | 0.18 | 0.24 | 0.09 | 0.20 | 0.15 | 0.17 | 0.14 | **0.16** | 17.6 |
| K3 | 0.09 | 0.15 | 0.13 | 0.16 | 0.11 | 0.14 | 0.15 | 0.15 | **0.14** | 45.0 | K1 | 0.09 | 0.15 | 0.24 | 0.12 | 0.17 | 0.16 | 0.17 | 0.12 | **0.15** | 17.0 |
| K1 | 0.11 | 0.25 | 0.10 | 0.09 | 0.11 | 0.19 | 0.14 | 0.04 | **0.13** | 39.8 | K2 | 0.06 | 0.13 | 0.22 | 0.10 | 0.15 | 0.14 | 0.15 | 0.10 | **0.13** | 14.8 |
| K4 | 0.10 | 0.09 | 0.06 | 0.10 | 0.06 | 0.07 | 0.10 | 0.02 | **0.07** | 30.4 | K4 | 0.14 | 0.09 | 0.13 | 0.04 | 0.08 | 0.11 | 0.21 | 0.07 | **0.11** | 14.2 |
| K7 | 0.00 | 0.03 | 0.03 | 0.03 | 0.03 | 0.02 | 0.04 | 0.12 | **0.04** | 6.7 | K7 | 0.05 | 0.11 | 0.12 | 0.04 | 0.14 | 0.11 | 0.14 | 0.13 | **0.11** | 13.0 |
| K9 | 0.00 | 0.02 | 0.05 | 0.08 | 0.03 | 0.03 | 0.04 | 0.02 | **0.03** | 5.7 | K9 | 0.03 | 0.10 | 0.07 | 0.04 | 0.04 | 0.04 | 0.03 | 0.08 | **0.06** | 6.2 |
| K5 | 0.00 | 0.02 | 0.05 | 0.05 | 0.03 | 0.02 | 0.04 | 0.03 | **0.03** | 5.0 | K6 | 0.03 | 0.07 | 0.06 | 0.02 | 0.06 | 0.05 | 0.04 | 0.08 | **0.05** | 5.3 |
| K8 | 0.00 | 0.02 | 0.02 | 0.05 | 0.03 | 0.02 | 0.03 | 0.03 | **0.03** | 4.9 | K8 | 0.02 | 0.04 | 0.05 | 0.02 | 0.03 | 0.03 | 0.03 | 0.09 | **0.04** | 4.3 |
| K6 | 0.00 | 0.01 | 0.02 | 0.05 | 0.03 | 0.01 | 0.02 | 0.01 | **0.02** | 2.7 | K5 | 0.03 | 0.04 | 0.05 | 0.04 | 0.04 | 0.04 | 0.03 | 0.03 | **0.04** | 4.0 |

In summary, the class separability assessment indicated that the use of the full Kennaugh-Matrix is more beneficial than use of the model-based (power) Decompositions, Eigen-decompositions, or the intensities of the polarimetric channels. L-Band, followed by C-Band and X-Band showed the best separation concerning the different wavelengths. Using the PolSAR feature spaces was most beneficial for the separation of the land cover classes OL, BS, BW and WT.

## 4. Discussion

Correlation analyses of the PolSAR features indicated that the quad-polarized Kennaugh Matrix elements K0, K1, K3, K4 and K7 were highly correlated with corresponding elements of the dual-polarized Kennaugh Matrices. As the dual-polarized Kennaugh matrix is a submatrix of the full-polarized Kennaugh Matrix generated out of these elements, the elements are therefore interchangeable and the dual-polarized data provide a substitute of the full quad-polarized data, at least for the tundra land cover investigated in this research. Contrary to this, the quad-polarized Kennaugh Matrix elements K5 and K6 were less correlated to the corresponding elements of the cross-polarized Kennaugh Matrix. Nevertheless, there are still benefits associated with the Kennaugh Matrix, since all kinds of PolSAR data can be stored, processed and analyzed in the same manner. It also provides a unified framework without any loss of information, and the capacity to interpret decomposed elements in a coherent and incoherent way since any other incoherent or coherent scattering matrix can be derived if necessary [31,42]. The Stokes coordinate system used for the definition of the Kennaugh Matrix seems to offer an appropriate approach to characterize the environment investigated in this research.

For the examined tundra land cover of the Tuktoyaktuk Peninsula, it was further shown that the elements of the Two Component Decompositions of [14] and [30] were highly correlated with the corresponding elements of the Yamaguchi Decomposition and—with lower significance—volume scattering, and HV intensity, respectively. Thus, the HH/VV-polarized data provide crucial information for describing the land covers considered in this research. As pointed out by [14], the correspondence of these quad- and HH/VV-polarized decomposition features is a function of the presence and influence of volume scattering processes, relative to contributions from the ground. Thus, due to the relatively short stature (height) of tundra vegetation, the observed correlations were high due to a lack of a significant volume scattering component. Further, the correlations between the features were generally higher at L-Band, compared to features at C-Band. This can be attributed to the longer wavelength of the ALOS and ALOS-2 sensors, and the absence/weakness of random scattering processes as the penetration depth is higher and volume scattering is less likely (thus the volume component is small relative to surface scattering).

The backscatter characteristics of the tundra land cover classes were examined via box- and scatterplots of the individual PolSAR features. It was shown that X-, C- and L-Band data exhibit distinct scattering characteristics for the different land cover classes. Results indicate that the L-Band data were more sensitive to the bare ground classes; thus, it is better suited to investigate and monitor ground properties, e.g., soil moisture, or the surface heave and subsidence (via InSAR) caused by the freezing and thawing of the active layer (compare [17,20,21]); especially in sites dominated by shrubs. In contrast, use of short wavelengths (X- and C-Band) is beneficial for characterizing tundra and wetland vegetation. This observation is in accordance with other studies [9,12,15].

It is worth noting the clear distinction of the land cover class: driftwood accumulation (BW) in the L-Band data. The coverage of BW is characterized by non-vegetated, dead woody debris, and frequently such accumulations exhibit a very high surface roughness, since dead wood and stems pile up more than a meter high (compare Figure 2c). Even though this should be a clearly visible target, and distinct feature in the PolSAR data, the position of BW is less clear in the feature spaces of X- and C-Band compared to the position of BW in the L-Band feature space. For BW the highest HH and HV intensity values (derived from ALOS in 2010) were found among all land cover classes. As well, the scattering from this type of coverage was characterized by high volume scattering and double bounce intensities of the Yamaguchi decomposition (derived from ALOS-2 in 2016) at L-Band. The dielectric and geometric properties of the driftwood accumulations facilitate high intensity scattering at L-Band, thus this type of coverage is a "rough" target at L-band but not in C- and X-Band (i.e., because the logs are much larger than incident C- and X-Band microwaves).

Even though the L-Band data showed limited value for characterizing the land cover classes HT, ST and WT using a single feature, the ALOS-2 quad-polarized and HH/VV-polarized data offered the feature space with the highest class separability; as indicated by the Transformed Divergence (TD) and squared Jefferys Matusita Distance (JD). However, since the ALOS-2 data were acquired at a steeper incidence angle and with a delay of six years, a true comparability of these results cannot be guaranteed. These results are therefore surprising, since one would assume a change of the land cover over time and an increasing dissimilarity between the reference and the PolSAR measurement with increasing temporal difference. Still, the ALOS HH/HV-polarized data acquired in 2010 showed a fairly good separability (Rank 17 in JD, Rank 18 in TD), and the data were observed to be more valuable for class separation than the C-Band (Rank 23 in JD, Rank 26 in TD), or X-Band (Rank 20 in TD, Rank 16 in JD) HH/HV-polarized data.

All separability features indicated that the Kennaugh Matrix was the most favorable feature space among all examined decompositions, which is in accordance with the expectation that full PolSAR information is better suited for class separability than is available via Entropy/Alpha, or the Two/Three Component Decomposition models, for instance. Among the Model-based Decompositions, the Yamaguchi Decomposition of quad-polarized data exceeded the separability offered by the Two Component Decomposition models. Thus, cross-polarized information is important for class discrimination, even though volume scattering processes play a minor role for the tundra environment investigated. Thus, perhaps differences in roughness/geometry play a more important role.

## 5. Conclusions

Results from this analysis indicate that the quad-polarized Kennaugh Matrix elements K0, K1, K3, K4 and K7 were highly correlated with corresponding elements of the dual-polarized Kennaugh Matrices; therefore, to a certain extent, dual-polarized data provide a useful substitute for the full quad-polarized data. The Kennaugh Matrix offers a unified framework to store, process and analyze PolSAR data in the same manner, and the Kennaugh elements offer comparable information from dual- or quad-polarized data. Thus, there is nearly no difference between the two acquisitions modes when using Kennaugh elements.

Among the investigated Model-based Decompositions and the Eigen-decompositions the features of the Two Component Decompositions models of [14] (based on HH/VV dual-polarized data) were highly correlated with the corresponding elements of the Yamaguchi Decomposition (based on quad-polarized data). Independent of the wavelength and polarization mode, the Eigen-decomposition features Entropy and the Alpha scattering angles were highly correlated and of less value for class separation. Therefore, this approach does not seem suitable for this low depolarizing as well as low entropy environment.

The X-, C- and L-Band data exhibit distinct scattering characteristics for the different land cover classes. The PolSAR data of all wavelengths are sensitive to the land cover classes: open water (OL), sand (BS) and mixed sediment (BM); L-Band data were most sensitive to the BW; X- and C-Band features were most sensitive to the inundated low-lying tundra (WI) and wetland WT, and herb dominated tundra (HT) and shrub dominated tundra (ST). The use of shorter wavelengths (X- and C-Band) is beneficial for characterizing wetland vegetation. The L-Band data exhibited the differences of the bare ground classes BS, BM and BW best. Thus, in accordance to previous studies L-Band data are favorable for InSAR applications in this region, due to the observed distinct surface scattering and the low volume scattering contribution. In contrast, C- and X-Band data are favorable for the characterization of the tundra land cover due to the observed sensitivity of the cross- and co-polarized information to tundra vegetation.

Nevertheless, the assessment of the class separability pointed out that PolSAR data of any wavelength—also of L-Band—were valuable for class separation and PolSAR information is beneficial for class discrimination. The results showed that quad-polarized data of ALOS-2 and R-2 offered the best separation of the land cover classes, followed by the HH/VV-, HH/HV- or VV/VH-polarized data of ALOS-2, R-2 and TSX. Further, full PolSAR information is better suited for class separation than less diverse polarimetric feature spaces, like all dual-polarimetric measurements (HH/VV, HH/HV or VV/VH). The Kennaugh Matrices offered the highest class separability among the investigated decompositions, and among the ten elements of the quad-polarized C- and L-Band Kennaugh Matrix the elements K0, K1, K2, K3 and K4 were found to be most valuable for class discrimination. This also indicates that the phase-relation between HH and VV (K3, K4) provides crucial information for separating the investigated tundra land cover classes, since it contains the distinction of surface from diplane scattering. Further, the intensity-based information of the elements K0, K1 and K2, which explain the diagonal elements of the Coherency Matrix, are favorable for class discrimination.

In light of the results presented in this manuscript, future work should focus on investigating the combined use of short- and long-wave PolSAR data, e.g., of C-/X-Band and L-Band. It is anticipated that such multi-frequency data will provide complementary information useful for accurate classification and the description of land surface parameters, as well as biophysical parameters of the tundra vegetation. In this context, the combination of PolSAR information via a multi-sensor approach seems very promising, since it will combine dielectric, and geometrical properties of the targets.

An interesting future question will be to also address the use of hybrid-polarimetric/ compact-polarimetric data that can be synthetically generated from quad-polarimetric data, also using the Kennaugh Matrix approach [31]. The question will be how such data perform compared to quad-, or dual-polarized data.

As well, the potential for land cover classification should be addressed, e.g., via the Random-Forest approach that was shown to provide an interesting classification framework also for PolSAR data [16,51]. In this context, upcoming studies should further acknowledge if the Random-Forest approach is appropriate and essential for a successful PolSAR classification.

The inclusion of Sentinel-1 C-Band PolSAR data is another option, as the Interferometric Wide-Swath mode provides large spatial coverage at high spatial resolutions, and the planned continuity of the Sentinel SAR systems will offer the capacity to support long term monitoring and consistent remote observations of Arctic land covers. However, as shown in this study the VV/VH polarization mode, employed by Sentinel-1 over most parts of the Canadian Arctic, seems less suited for characterizing of the tundra land cover classes; thus the use of a multi-frequency or multi-sensor approach is advisable.

In summary, the SAR data of all wavelengths—also of the L-Band—were shown to provide important information about the tundra environment and utilization of such remotely sensed information is strongly recommended. PolSAR data provide unique information on dielectric, and geometrical properties that can help to increase the information space. Whenever possible dual- or quad-polarized data should be used, as polarimetry was shown to be of high value and importance.

**Acknowledgments:** The authors like to thank Achim Roth (DLR) and Roland Baumhauer (University of Wuerzburg) for the helpful discussions and support. Further, we like to thank Jason Duffe and Blair E. Kennedy (NWRC & Carleton University Ottawa) for the organization of the field trip and the support during the field work. TerraSAR-X imagery is shown under permission of German Remote Sensing Data Center (DFD): Related Proposals COA1736 and COA1144. RADARSAT-2 imagery is shown under permission of MacDonald Dettwiler and Associates (MDA)—Related Multi-User Request Form MDA GSI-Ref.-Number: CG0061(2)-12-2011. ALOS and ALOS-2 imagery was obtained via the JAXA Proposal "Derivation of Environmental Parameters of Arctic Tundra Landscapes from Radar Remote Sensing"—PI No. 1186: Dr. Jennifer Sobiech-Wolf (Alfred Wegener Institute for Polar and Marin Research). This publication was funded by the German Research Foundation (DFG) in the funding program "Open Access Publishing" and by the University of Wuerzburg. The authors would like to thank the anonymous reviewers for their helpful and constructive comments.

**Author Contributions:** Tobias Ullmann, Sarah N. Banks and Andreas Schmitt conceived and designed the experiments; Tobias Ullmann and Sarah N. Banks performed the experiments; all authors analyzed the data and wrote the paper.

**Conflicts of Interest:** The authors declare no conflict of interest.

## Appendix A

Definition of the Kennaugh Matrix and its elements for quad-polarized (A1), HH/VV-polarized (A2) and HH/HV- or VV/VH-polarized data (A3) is reported in the following and based on [31].

$S_{XX}$ and $S_{YY}$ refer to the complex signals of the co-polarized channels. $S_{XY}$ refers to the complex signal of a cross-polarized channel.

$$\mathbf{K}_{quad} = \begin{bmatrix} K0 & K4 & K5 & K6 \\ K4 & K1 & K9 & K8 \\ K5 & K9 & K2 & K7 \\ K6 & K8 & K7 & K3 \end{bmatrix}, \text{ with}$$

$$K0 = 0.5\left(|S_{XX}|^2 + 2|S_{XY}|^2 + |S_{YY}|^2\right)$$
$$K1 = 0.5\left(|S_{XX}|^2 - |S_{XY}|^2 - |S_{YX}|^2 + |S_{YY}|^2\right)$$
$$K2 = \Re(S_{XX}S_{YY}^* - S_{XY}S_{XY}^*)$$
$$K3 = -\Re(S_{XX}S_{YY}^* - S_{XY}^*S_{XY})$$
$$K4 = 0.5\left(|S_{XX}|^2 - |S_{XY}|^2 + |S_{XY}|^2 - |S_{YY}|^2\right)$$
$$K5 = \Re(S_{XX}S_{XY}^* + S_{XY}S_{YY}^*)$$
$$K6 = \Im(S_{XX}S_{XY}^* + S_{XY}S_{YY}^*)$$
$$K7 = \Im(S_{XX}S_{YY}^* + S_{XY}S_{XY}^*)$$
$$K8 = \Im(S_{XX}S_{XY}^* - S_{XY}S_{YY}^*)$$
$$K9 = \Re(S_{XX}S_{XY}^* - S_{XY}S_{YY}^*)$$

(A1)

$$\mathbf{K}_{HH/VV.} = \begin{bmatrix} K0 & K4 & 0 & 0 \\ K4 & 0 & 0 & 0 \\ 0 & 0 & K2 & K7 \\ 0 & 0 & K7 & 0 \end{bmatrix}, \text{ with}$$

$$K0 = 0.5\left(|S_{XX}|^2 + |S_{YY}|^2\right)$$
$$K4 = 0.5\left(|S_{XX}|^2 - |S_{YY}|^2\right)$$
$$K3 = -\Re(S_{XX}S_{YY}^*)$$
$$K7 = \Im(S_{XX}S_{YY}^*)$$

(A2)

$$\mathbf{K}_{XHH/HV \ or \ VV/VH.} = \begin{bmatrix} K0 & 0 & K5 & K6 \\ 0 & K1 & 0 & 0 \\ K5 & 0 & 0 & 0 \\ K6 & 0 & 0 & 0 \end{bmatrix}, \text{ with}$$

$$K0 = 0.5\left(|S_{XX}|^2 + 2|S_{XY}|^2\right)$$
$$K1 = 0.5\left(|S_{XX}|^2 - 2|S_{XY}|^2\right)$$
$$K5 = \Re(S_{XX}S_{XY}^*)$$
$$K6 = \Im(S_{XX}S_{XY}^*)$$

(A3)

Conversion of Kennaugh Matrix to $3 \times 3$ Coherency Matrix (**T**) of quad-polarized (A4) and $2 \times 2$ T of HH/VV-polarized data (A5) is defined according to [31,42] as:

$$\mathbf{T}_{quad} = \begin{bmatrix} T_{11} & T_{12} & T_{13} \\ T_{21} & T_{22} & T_{23} \\ T_{31} & T_{22} & T_{33} \end{bmatrix} =$$
$$= \begin{bmatrix} 0.5(K0+K1+K2-K3) & K4-iK7 & K5+iK8 \\ K4+iK7 & 0.5(K0+K1-K2+K3) & K9+iK6 \\ K5-iK8 & K9-iK6 & 0.5(K0-K1+K2+K3) \end{bmatrix} =$$

(A4)

$$\mathbf{T}_{HH/VV} = \begin{bmatrix} T_{11} & T_{12} \\ T_{21} & T_{22} \end{bmatrix} = \begin{bmatrix} K0-K3 & K4-iK7 \\ K4+iK7 & K0+K3 \end{bmatrix}$$

(A5)

## References

1. Jeffries, M.; Morris, K.; Liston, G. A method to determine lake depth and water availability on the North Slope of Alaska with spaceborne imaging radar and numerical ice growth modelling. *ARCTIC* **1996**, *49*, 367–374. [CrossRef]
2. Kozlenko, N.; Jeffries, M. Bathymetric mapping of shallow water in thaw lakes on the North Slope of Alaska with spaceborne imaging radar. *ARCTIC* **2000**, *53*, 306–316. [CrossRef]
3. Hall-Atkinson, C.; Smith, L.C. Delineation of delta ecozones using interferometric SAR phase coherence Mackenzie River Delta, N.W.T., Canada. *Remote Sens. Environ.* **2001**, *78*, 229–238. [CrossRef]
4. Hugenholtz, C.; Sanden, V.-D.J. *Polarimetric SAR for Geomorphic Mapping in the Intertidal Zone, Minas Basinm Bay of Fundy, Nova Scotia*; Natural Resources Canada & Canadian Center for Remote Sensing: Ottawa, ON, Canada, 2011; pp. 1–28.
5. May, I.; Ludwig, R.; Bernier, M. Using TerraSAR-X imagery for the monitoring of permafrost dynamics in Northern Quebec. In Proceedings of the 4th TerraSAR-X Science Team Meeting, Oberpfaffenhofen, Germany, 14–16 February 2011; pp. 1–8.
6. Banks, S.N.; King, D.J.; Merzouki, A.; Duffe, J.; Solomon, S. Assessing Radarsat-2 polarimetric SAR for mapping shoreline cleanup and assessment technique (SCAT) classes in the Canadian. In Proceedings of the 32nd Canadian Symposium on Remote Sensing, Sherbrooke, QC, Canada, 13–16 June 2011; pp. 1–8.
7. Banks, S.; King, D.; Merzouki, A.; Duffe, J. Assessing RADARSAT-2 for mapping shoreline cleanup and assessment technique (SCAT) classes in the Canadian Arctic. *Can. J. Remote Sens.* **2014**, *40*, 243–267. [CrossRef]
8. Sobiech, J.; Boike, J.; Dierking, W. Observation of melt onset in an arctic tundra landscape using high resolution TerraSAR-X and RADARSAT-2 data. In Proceedings of the 2012 IEEE International Geoscience and Remote Sensing Symposium (IGARSS), Munich, Germany, 22–27 July 2012; pp. 3552–3555.
9. Regmi, P.; Grosse, G.; Jones, M.; Jones, M.; Anthony, K. Characterizing post-drainage succession in thermokarst lake basins on the Seward Peninsula, Alaska with TerraSAR-X backscatter and Landsat-based NDVI data. *Remote Sens.* **2012**, *4*, 3741–3765. [CrossRef]
10. Jagdhuber, T.; Stockamp, J.; Hajnsek, I.; Ludwig, R. Identification of Soil Freezing and Thawing States Using SAR Polarimetry at C-band. *Remote Sens.* **2014**, *6*, 2008–2023. [CrossRef]
11. Ullmann, T.; Schmitt, A.; Roth, A.; Duffe, J.; Dech, S.; Hubberten, H.-W.; Baumhauer, R. Land Cover Characterization and Classification of Arctic Tundra Environments by Means of Polarized Synthetic Aperture X- and C-Band Radar (PolSAR) and Landsat 8 Multispectral Imagery—Richards Island, Canada. *Remote Sens.* **2014**, *6*, 8565–8593. [CrossRef]
12. Collingwood, A.; Treitz, P.; Charbonneau, F.; Atkinson, D. Artificial neural network modelling of high arctic phytomass using synthetic aperture radar and multispectral data. *Remote Sens.* **2014**, *6*, 2134–2153. [CrossRef]
13. Banks, S.; Millard, K.; Pasher, J.; Richardson, M.; Wang, H.; Duffe, J. Assessing the Potential to Operationalize Shoreline Sensitivity Mapping: Classifying Multiple Wide Fine Quadrature Polarized RADARSAT-2 and Landsat 5 Scenes with a Single Random Forest Model. *Remote Sens.* **2015**, *7*, 13528–13563. [CrossRef]
14. Ullmann, T.; Schmitt, A.; Jagdhuber, T. Two Component Decomposition of Dual Polarimetric HH/VV SAR Data: Case Study for the Tundra Environment of the Mackenzie Delta Region, Canada. *Remote Sens.* **2016**, *8*, 1027. [CrossRef]
15. Widhalm, B.; Bartsch, A.; Leibmann, M.; Khomutov, A. Active-layer thickness estimation from X-band SAR backscatter intensity. *Cryosphere* **2017**, *11*, 483–496.
16. Brisco, B.; Short, N.; Budkewitsch, P.; Murnaghan, K.; Charbonneau, F. SAR interferometry and polarimetry for mapping and monitoring permafrost in Canada. In Proceedings of the 4th International Workshop on Science and Applications of SAR Polarimetry and Polarimetric Interferometry (PolInSAR 2009), Frascati, Italy, 26–30 January 2009; pp. 1–4.
17. Short, N.; Brisco, B.; Couture, N.; Pollard, W.; Murnaghan, K.; Budkewitsch, P. A comparison of TerraSAR-X, RADARSAT-2 and ALOS-PALSAR interferometry for monitoring permafrost environments, case study from Herschel Island, Canada. *Remote Sens. Environ.* **2011**, *115*, 3491–3506. [CrossRef]
18. Short, N.; LeBlanc, A.-M.; Sladen, W.; Oldenborger, G.; Mathon-Dufour, V.; Brisco, B. RADARSAT-2 D-InSAR for ground displacement in permafrost terrain, validation from Iqaluit Airport, Baffin Island, Canada. *Remote Sens. Environ.* **2014**, *141*, 40–51. [CrossRef]

19. Schaefer, T. Remotely Sensed Active Layer Thickness (ReSALT) at Barrow, Alaska Using Interferometric Synthetic Aperture Radar. *Remote Sens.* **2015**, *7*, 3735–3759. [CrossRef]

20. Iwahana, G.; Uchida, M.; Liu, L.; Gong, W.; Meyer, F.J.; Guritz, R.; Yamanokuchi, T.; Hinzman, L. InSAR Detection and Field Evidence for Thermokarst after a Tundra Wildfire, Using ALOS-PALSAR. *Remote Sens.* **2016**, *8*, 218. [CrossRef]

21. Jia, Y.; Kim, J.-W.; Shum, C.K.; Lu, Z.; Ding, X.; Zhang, L.; Erkan, K.; Kuo, C.-Y.; Shang, K.; Tseng, K.-H.; et al. Characterization of Active Layer Thickening Rate over the Northern Qinghai-Tibetan Plateau Permafrost Region Using ALOS Interferometric Synthetic Aperture Radar Data, 2007–2009. *Remote Sens.* **2017**, *9*, 84. [CrossRef]

22. Larsen, J.; Anisimov, O.; Constable, A.; Hollowed, A.B.; Maynard, N.; Prestrud, P.; Prowse, T.; Stone, J. Polar Regions. In *Climate Change 2014: Impacts, Adaptation, and Vulnerability—Part B: Regional Aspects. Contribution of Working Group II to the Fifth Assessment Report of the Intergovernmental Panel on Climate Change*; Barros, V.R., Field, C.B., Dokken, D.J., Mastrandrea, M.D., Mach, K.J., Bilir, T.E., Chatterjee, M., Ebi, K.L., Estrada, Y.O., Genova, R.C., et al., Eds.; Cambridge University Press: Cambridge, UK, 2014; pp. 1567–1612.

23. Lawrence, D.M.; Slater, A.G.; Tomas, R.A.; Holland, M.M.; Deser, C. Accelerated Arctic land warming and permafrost degradation during rapid sea ice loss. *Geophys. Res. Lett.* **2008**, *35*, L11506. [CrossRef]

24. Romanovsky, V.E.; Smith, S.L.; Christiansen, H.H. Permafrost thermal state in the polar northern hemisphere during the international polar year 2007–2009: A synthesis. *Permafr. Periglac. Proc.* **2010**, *21*, 106–116. [CrossRef]

25. Cloude, S.R.; Pottier, E. A review of target decomposition theorems in radar Polarimetry. *IEEE Trans. Geosci. Remote Sens.* **1996**, *34*, 498–518. [CrossRef]

26. Cloude, S.R. The Dualpol Entropy/Alpha decomposition: A PALSAR case study. In Proceedings of the 3th International Workshop on Science and Applications of SAR Polarimetry and Polarimetric Interferometry (PolInSAR), Frascati, Italy, 22–26 January 2007.

27. Yamaguchi, Y.; Yajima, Y.; Yamada, H. A four-component decomposition of POLSAR images based on the Coherency Matrix. *IEEE Trans. Geosci. Remote Sens.* **2006**, *3*, 292–296. [CrossRef]

28. Freeman, A.; Durden, S. A three-component scattering model for polarimetric SAR data. *IEEE Trans. Geosci. Remote Sens.* **1998**, *36*, 963–973. [CrossRef]

29. Touzi, R.; Goze, S.; Le Toan, T.; Lopes, A.; Mougin, E. Polarimetric discriminators for SAR images. *IEEE Geosci. Remote Sens.* **1992**, *30*, 973–980. [CrossRef]

30. Jagdhuber, T.; Hajnsek, I.; Caputo, M.; Papathanassiou, K.P. Soil Moisture Estimation Using Dual-Polarimetric Coherent (HH/VV) TerraSAR-X and TanDEM-X Data. In Proceedings of the TSX/TDX Science Meeting, Oberpfaffenhofen, Germany, 10–14 June 2013.

31. Schmitt, A.; Wendleder, A.; Hinz, S. The Kennaugh element framework for multi-scale, multi-polarized, multi-temporal and multi-frequency SAR image preparation. *ISPRS J. Photogramm. Remote Sens.* **2015**, *102*, 122–139. [CrossRef]

32. Ecological Stratification Working Group (Canada); Center for Land and Biological Resources Research (Canada); State of the Environment Directorate, Canada. *A National Ecological Framework for Canada*; Centre for Land and Biological Resources Research, Research Branch, Agriculture and Agri-Food Canada: Ottawa, ON, Canada, 1996; pp. 1–132.

33. Burn, C.R.; Kokelj, S.V. The environment and permafrost of the Mackenzie Delta Area. *Permafr. Periglac. Proc.* **2009**, *20*, 83–105. [CrossRef]

34. NWT-Geomatics. Northwest Territories (NWT) Centre for Geomatics. 2016. Available online: http:geomatics.gov.nt.ca (accessed on 15 April 2017).

35. Government of Canada; Natural Resources Canada; Earth Sciences Sector; Canada Centre for Mapping and Earth Observation. *GeoBase-Land Cover, Circa 2000 Vector Data Product Specifications*; Centre for Topographic Information Earth Sciences Sector Natural Resources Canada: Ottawa, ON, Canada, 2009; pp. 1–21.

36. Corns, I.G.W. Arctic plant communities east of the Mackenzie Delta. *Can. J. Bot.* **1974**, *52*, 1731–1745. [CrossRef]

37. Moffat, N.D.; Lantz, T.C.; Fraser, R.H.; Olthof, I. Recent Vegetation Change (1980–2013) in the Tundra Ecosystems of the Tuktoyaktuk Coastlands, NWT, Canada. *Arct. Antarct. Alp. Res.* **2016**, *48*, 581–597. [CrossRef]

38. Lantz, T.C.; Marsh, P.; Kokelj, S.V. Recent Shrub Proliferation in the Mackenzie Delta Uplands and Microclimatic Implications. *Ecosystems* **2013**, *16*, 47–59. [CrossRef]

39. Richards, J.A. *Remote Sensing with Imaging Radar*; Springer: Berlin/Heidelberg, Germany, 2009; pp. 1–361.

40. Ullmann, T.; Büdel, C.; Baumhauer, R. Characterization of Arctic Surface Morphology by Means of Intermediated TanDEM-X Digital Elevation Model Data. *Z. Geomorphol.* **2017**, *61*, 3–25. [CrossRef]

41. Guissard, A. Mueller and Kennaugh matrices in radar polarimetry. *IEEE Trans. Geosci. Remote Sens.* **1994**, *32*, 590–597. [CrossRef]

42. Cloude, S.R. *Polarisation—Applications in Remote Sensing*; Oxford University Press: Oxford, UK, 2009; pp. 1–453.

43. Lee, J.-S.; Pottier, E. Introduction to the Polarimetric Target Decomposition Concept. In *Polarimetric Radar Imaging: From Basics to Applications*; CRC Press: Boca Raton, FL, USA, 2009; pp. 1–422.

44. Jagdhuber, T.; Hajnsek, I.; Caputo, M.; Papathanassiou, K.P. Dual-Polarimetry for soil moisture inversion at X-Band. In Proceedings of the EUSAR, Berlin, Germany, 3–5 June 2014; pp. 1–4.

45. Jensen, J.R. *Introductory Digital Image Processing: A Remote Sensing Perspective*, 2nd ed.; Prentice Hall PTR: Upper Saddle River, NJ, USA, 1995; pp. 1–544.

46. Swain, P.H. *A Result from Studies of Transformed Divergence*; LARS Technical Reports; Laboratory Applications of Remote Sensing, Purdue University: West Lafayette, IN, USA, 1973; Volume 42, pp. 1–5.

47. Bhattacharyya, A. On a measure of divergence between two statistical populations defined by their probability distributions. *Bull. Calcutta Math. Soc.* **1943**, *35*, 99–109.

48. Mausel, P.W.; Kramber, W.J.; Lee, J.K. Optimal band selection for supervised classification of multispectral data. *Photogramm. Eng. Remote Sens.* **1990**, *56*, 55–60.

49. Swain, P.H.; Davis, S.M. *Remote Sensing: The Quantitative Approach*; McGraw Hill Book Company: New York, NY, USA, 1978.

50. Mitsunobu, S.; Kazuo, O.; Chan-Su, Y. On the eigenvalue analysis using HH-VV dual-polarization SAR data and its applications to monitoring of coastal oceans. In Proceedings of the SPIE Conference on Ocean Sensing and Monitoring V, Baltimore, MD, USA, 30 April–1 May 2013; p. 8724.

51. Heine, I.; Jagdhuber, T.; Itzerott, S. Classification and Monitoring of Reed Belts Using Dual-Polarimetric TerraSAR-X Time Series. *Remote Sens.* **2016**, *8*, 552. [CrossRef]

*applied*
*sciences*

MDPI

Article

# Synergetic of PALSAR-2 and Sentinel-1A SAR Polarimetry for Retrieving Aboveground Biomass in Dipterocarp Forest of Malaysia

**Hamdan Omar \*, Muhamad Afizzul Misman and Abd Rahman Kassim**

Geoinformation Programme, Division of Forestry and Environment, Forest Research Institute Malaysia, Selangor 52109, Malaysia; afizzul@frim.gov.my (M.A.M.); rahmank@frim.frim.gov.my (A.R.K.)
\* Correspondence: hamdanomar@frim.gov.my; Tel.: +60-3-6279-7200

Academic Editors: Juan Manuel Lopez-Sanchez and Carlos López-Martínez
Received: 15 April 2017; Accepted: 17 June 2017; Published: 30 June 2017

**Abstract:** Space borne synthetic aperture radar (SAR) data have become one of the primary sources for aboveground biomass (AGB) estimation of forests. However, studies have indicated that limitations occur when a single sensor system is employed, especially in tropical forests. Hence, there is potential for improving estimates if two or more different sensor systems are used. Studies on integrating multiple sensor systems for estimation of AGB over Malaysia's tropical forests are scarce. This study investigated the use of PALSAR-2 L-band and Sentinel-1A C-band SAR polarizations to estimates the AGB over 5.25 million ha of the lowland, hill, and upper hill forests in Peninsular Malaysia. Polarized images, i.e., HH–HV from PALSAR-2 and VV–VH from Sentinel-1A have been utilized to produce several variables for predictions of the AGB. Simple linear and multiple linear regression analysis was performed to identify the best predictor. The study concluded that although limitations exist in the estimates, the combination of all polarizations from both PALSAR-2 and Sentiel-1A SAR data able to increase the accuracy and reduced the root means square error (RMSE) up to 14 Mg ha$^{-1}$ compared to the estimation resulted from single polarization. A spatially distributed map of AGB reported the total AGB within the study area was about 1.82 trillion Mg of the year 2016.

**Keywords:** aboveground biomass; tropical forest; microwave sensor system

## 1. Introduction

Aboveground biomass (AGB) includes all vegetation above the ground (i.e., stems, branches, bark, seeds, flowers, and foliage of live plants) and approximately 50% of its composition is carbon [1]. AGB usually measures in metric tons of dry matter per hectare (e.g., t ha$^{-1}$ or Mg ha$^{-1}$) or in metric tons of carbon per hectare (e.g., t C ha$^{-1}$ or Mg C ha$^{-1}$). The United Nations Framework Convention on Climate Change (UNFCCC) identified it as an Essential Climate Variable (ECV). Therefore, accurate information on biomass stock in world forests is necessary to reduce uncertainties and to fill the knowledge gaps of the climate system [2]. Further strong impetus to improve methods for measuring global biomass comes from the reduction of emissions due to deforestation and forest degradation (REDD) mechanism, which was introduced in the UNFCCC Committee of the Parties (COP-13) Bali Action Plan. REDD which is now popular with REDD+ (with additional elements of carbon stock enhancement and biodiversity conservation) is dedicated to the developing countries around the world including Malaysia. Its implementation relies fundamentally on systems to assess available carbon stock and monitor changes due to loss of biomass from deforestation and forest degradation [3], which are amalgamated in a system called monitoring, reporting, and verification (MRV).

Remote sensing has been recognized as one of the primary spatial inputs for this process [4–6]. Satellite remote sensing technologies are currently widely tested and suggested as a tool in REDD+

MRV. Along with scientific programs and field tests, there is also a debate as to the overall feasibility and cost–benefit ratio of remote sensing approaches, depending on the wide range of ecosystem and land use conditions as well as the range of approaches to carbon credit accounting [7].

In many parts of the world, especially in tropical region, the frequent cloud conditions often restrain the acquisition of high-quality remotely sensed data by optical sensors. The acquisition of cloud-free, wall-to-wall optical satellite images in tropical countries is almost impossible [8]. Thus, SAR data become the only feasible way of acquiring remotely sensed data within a given timeframe because the SAR systems are independent of cloud coverage, weather, and light conditions. Due to this unique feature compared with optical sensor data, the SAR data have been used extensively in many fields, including forest-cover identification and mapping, discrimination of forest from other land covers, and forest biomass estimation.

Previous studies demonstrated that the L-band polarimetry backscatter tends to saturate at certain levels of biomass, and hence limits the accuracy of estimates [9–11]. However, the saturation level varies with the type and structure of the forest. It was demonstrated that the sensitivity of SAR polarimetry is depending on the structure, density, and tree elements (i.e., trunk/stem, branches, and leaves) of the forests [12]. Other than these issues, several other inter-related issues can affect the biomass estimations using remotely sensed data. These issues can be generalized into three major groups, which are (i) the natural conditions of the forest, (ii) the forest management system being practiced, and (iii) the technical issues related to the remote sensing system being used [13].

Short wavelength SAR sensors on board several satellites such as the Earth Resources Satellite (ERS-1), Radarsat, and Environmental Satellite (Envisat) have been used to quantify forest biomass. A number of studies have been conducted in relatively homogeneous or young forests, but the signal tends to saturate at low biomass (100–200 Mg ha$^{-1}$) [14–16]. However, L-band SAR has shown better potential in retrieving the biomass of forests, including those in the tropics [17–21]. Recently, there has been rising interest in integrating data from several SAR sensors and SAR with optical sensors to improve the accuracy of biomass estimates [22,23].

In Malaysia, there are limited studies on the applications of SAR for estimating biomass. Out of many studies conducted worldwide, very few have been done in Malaysia [11,24,25]. This indicates that the potential, limitations, and advances of L-band SAR in estimating tropical forest in Malaysia are not extensively explored. Methods of applying this SAR system are also scarcely exploited. Therefore, the objective of this study is to explore the synergy of SAR sensors, i.e., PALSAR-2 L-band and Sentinel-1A C-band for estimation of AGB in inland dry dipterocarp forest in Peninsular Malaysia. This study highlights and discusses advantages and limitations of this technique.

## 2. Materials and Methods

### 2.1. The Study Area

The study area comprised lowland, hill, and upper hill dipterocarp forests, which are categorized based on land altitude, i.e., <300, 300–750, and 750–1200 m, respectively. These forests are major, occupying about 5,257,395 ha or about 89% out of the total forested land (i.e., about 5.9 million ha) in Peninsular Malaysia. These forests occur within the entire Peninsular Malaysia, which has an extent between 1–7° latitude and 99–105° longitude. These forests embrace all the well-drained primary forests of the plains, undulating land, and foothills and hill terrain up to about 750 m altitude. Trees from the family of Dipterocarpaceae are dominant species, which make the forests major timber production areas in Peninsular Malaysia. Almost the entire area (i.e., 4.9 million ha) is categorized as Reserve Forest which is meant for production and protection. About 1.98 million ha have been allocated for protection forests in the form of national parks, wildlife sanctuaries, and nature reserves [26]. The most common tree species found in this forest come from the genera such as Shorea, Hopea, Dipterocarpus, Dryobalanops, Neobalacarpus, Anisoptera, and Vatica. The remaining forested land is comprised of peat swamp, mangrove, and montane forests. Figure 1 shows the distribution of major forest types in Peninsular Malaysia.

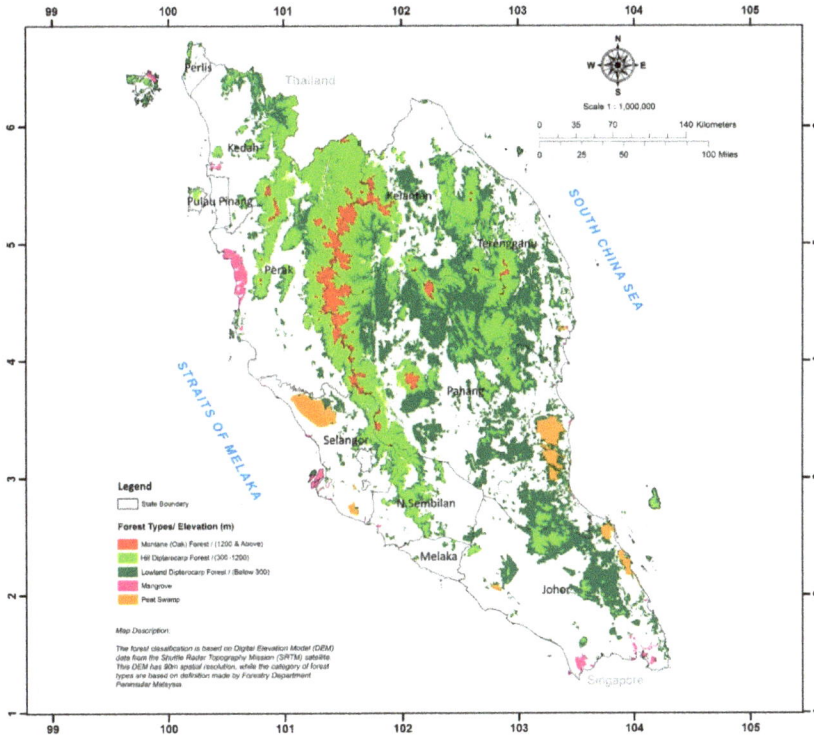

**Figure 1.** Forest types in Peninsular Malaysia.

## 2.2. Satellite Datasets

### 2.2.1. Satellite Images Acquisition

The satellite datasets that have been used in this study came from two satellites, which are; (i) Advanced Land Observing Satellite 2 (ALOS-2) that carries Phased Array type L-band Synthetic Aperture Radar-2 (PALSAR-2) on-board and (ii) Sentinel-1A that carries C-band imaging SAR sensor. Table 1 summarizes the properties of the data and Figure 2 shows both datasets that have been processed and used in this study.

ALOS-2 is the successor of the ALOS, but the structure of the new satellite is quite different from its predecessor. ALOS was launched in January 2006 and brought the Phased Array type L-band Synthetic Aperture Radar (PALSAR) on-board. After five years of observations, it stopped transmitting in April 2011. ALOS-2 was then launched on 24 May 2014, which carried the PALSAR-2 sensor. PALSAR-2 is currently operating and producing L-band SAR data, that has similar (with some advancements) characteristics with PALSAR. The data were acquired from Earth Observation Research Center (EORC) under Japan Aerospace Exploration Agency (JAXA). The data was acquired under the Kyoto and Carbon (K&C) Initiave, a research agreement between Forest Research Institute Malaysia (FRIM) and JAXA, whereby FRIM has special permission to access the PALSAR-2 product at all imaging modes and resolutions. JAXA also provides free access of PALSAR-2 mosaic product at 25-meter resolution for public which is available at http://www.eorc.jaxa.jp/ALOS/en/PALSAR_fnf/data/index.htm.

The SENTINEL-1 mission is the European Radar Observatory for the Copernicus joint initiative of the European Commission (EC) and the European Space Agency (ESA). The SENTINEL-1 was launched on 3 April 2014 and its mission operating in four exclusive imaging modes with different resolution (down to 5 m) and coverage (up to 400 km). It provides dual polarization capability, very short revisit times, and rapid product delivery. The Sentine-1A data was acquired in Level-1 Ground Range Detected (GRD) format so that radar cross-section of both distributed and point targets can be easily derived from the data. The data is available at https://scihub.copernicus.eu/dhus/#/home and free to download.

Another satellite data that was used in this study was the digital elevation model acquired from the Shuttle Radar Topography Mission (SRTM). This data was used to classify the forest into specified elevation categories, according to the type of forests. It was also used for radiometric terrain correction on both PALSAR-2 and Sentinel-1A images. These data are available at the US Geological Survey's EROS Data Center for download at http://srtm.usgs.gov/index.html.

**Table 1.** Summary of satellite images used in this study.

| Sensor | Wavelength | Date of Acquisition | Mode/Polarization | No. of Scene | Ground Resolution (m) | Incidence Angle (°) |
|---|---|---|---|---|---|---|
| PALSAR-2 | C-band (5.405 GHz) | Between March and June 2016 | Fine Beam Dual (FBD)/HH, HV | 52 | 6 | 29.1–46.0 |
| Sentinel-1A | L-band (1.270 GHz) | November 2016 | Interferometric Wide swath (IW) VV, VH | 8 | 9 | 37 |

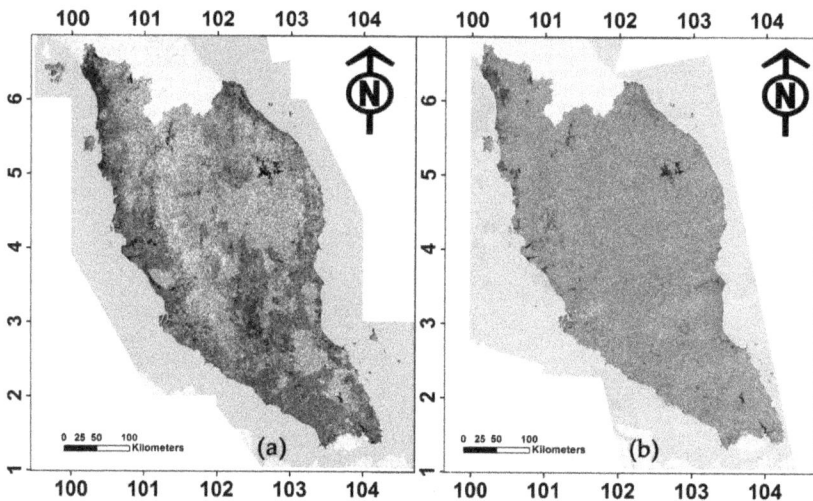

**Figure 2.** Mosaics of (**a**) PALSAR-2 and (**b**) Sentinel-1A over Peninsular Malaysia, displayed HV and VH polarizations, respectively.

### 2.2.2. Satellite Image Pre-Processing

PALSAR-2 images that were used in this study came in Level-1.5 product, which means the range and multilook azimuth compressed data is represented by amplitude data. The range coordinates were also converted from slant range to ground range, and map projection was performed.

Level-1 GRD consist of Sentinel-1A products consist of focused SAR data that has been detected, multi-looked and projected to ground range using an Earth ellipsoid model such as WGS84. The ellipsoid projection of the GRD products is corrected using the terrain height specified in the product general annotation. The terrain height used varies in azimuth but is constant in range.

Ground range coordinates are the slant range coordinates projected onto the ellipsoid of the Earth. Pixel values represent detected magnitude. Phase information is lost. The resulting product has approximately square resolution pixels and square pixel spacing with reduced speckle at a cost of reduced geometric resolution. In addition, the GRD products have thermal noise removed to improve the quality of the detected image.

Other than these processes, they are two important pre-processing stages, namely, speckle suppression and radiometric terrain correction. Spatial domain Lee Sigma filter with a kernel size of 7 × 7 pixels was used to remove speckle effect on the images. Digital Elevation Model (DEM) acquired from SRTM was used to minimize terrain shadowing effect on the images. The presence of this effect on SAR imagery was because the signal strengths were dependent on two variables, which were incidence angle and surface roughness or topography of terrain. If slope is facing the SAR transmitter, the signal will become stronger than the other side of the slope. Semi-empirical method was used for radiometric terrain correction [27] and the processes were performed in ENVI/IDL (Harris Corporation, Melbourne, Australia) following the same approach as found in Canty et al. [28]. This process was necessary to normalize both sides of the slopes and it minimized errors towards the end of AGB prediction.

### 2.2.3. Satellite Image Calibration

The objective of SAR calibration is to provide imagery in which the pixel values can be directly related to the radar backscatter of the scene. The PALSAR-2 image that was used in this study was built on a 16-bit data type and all pixels have digital numbers (DN) ranging from 0 to 65,535. These DNs however do not represent the radar signal of features or objects on the ground. Therefore, the DNs have to be converted to backscatter (i.e., the returned radar signals) known as Normalized Radar Cross Section (NRCS) and represented as $\sigma^0$ in decibels (dB). The equation that was used for the calculation of NRCS for PALSAR are slightly different from other sensors in that the usual sine term has already been included in the DN values. Thus, for the products stored at Level 1.5 and above, the equation for NRSC of any of the polarization component can be obtained by the following formula with single calibration factor (CF), which can be expressed as follows [29].

$$\sigma_{dB}^0 = 10 \cdot log_{10}\left(DN^2\right) - 83 \tag{1}$$

The Sentinel-1A product uses radiometric calibration look-up table (LUT) to do the calibration. This was performed on Sentinel Application Platform (SNAP) tool, a software that was designed specifically for Sentinel-1 products processing and it available for free at http://step.esa.int/main/download/. The essential conversion of amplitude to $DN$ and from $DN$ to sigma nought were done automatically on SNAP and once the sigma nought values was obtained, the computation of backscatter ($\sigma_{dB}^0$) can be performed as

$$\sigma_{dB}^0 = 10 \cdot log_{10}\sigma^0 \tag{2}$$

### 2.3. Forest-Non-Forest Classification

Forest-non-forest (FNF) classification was performed on PALSAR-2 polarizations images to delineate the forests from other land cover. This process is critical to define the boundary of forests and to ensure that the estimated AGB did not include other types of vegetation. The reason was that the forests are often confused with rubber, teak, and other timber tree plantations, which are common in Peninsular Malaysia and they appear almost identical on both HH and HV polarizations. Instead of using only the original backscatter HH and HV polarizations, an attempt has been also made to derive other image variables derived from PALSAR HH and HV images. Image variables, namely (i) simple band ratio (HH/HV), (HV/HH), (ii) average (HH + HV/2), and (iii) square root of products ($\sqrt{(HH \times HV)}$) were produced.

The incorporation of texture measure also can improve classification of spatially distributed pixels on an image. Gray-level co-occurrence matrix (GLCM) uses a gray-tone spatial dependence matrix to calculate texture values. This is a matrix of relative frequencies with which pixel values occur in two neighboring processing windows separated by a specified distance and direction. For this purpose, texture has been defined as repeating pattern of local variations in image intensity which is too fine to be distinguished as separate class at the observed resolution. Thus, a connected set of pixels satisfying given gray-level properties which occur repeatedly in an image region constitute a textured region. A mean-type GLCM was applied to the original HH and HV polarizations to produce textured images with clearer definitions of the objects on the images [30].

These inputs were used for the FNF classification and the Maximum Likelihood Classifier algorithms with nearest neighbor technique was applied. The forests were then further classified into several forest types by using the DEM from SRTM.

*2.4. Forest Survey Data*

The sampling design in this study modified from the standard operating procedure (SOP) that was developed by Winrock International [31], which follows the Intergovernmental Panel on Climate Change (IPCC) standards [1]. A cluster comprises of four plots and the design is shown in Figure 3. The plot was designed in circular with smaller nests inside. The biggest nest measures 20 m in radius, followed by the smaller nests, measuring 12 and 4 m. The sizes of trees were measured according to the nest sizes, which is summarized in Table 2. Depending on the nest size, it indicates that not all stands were measured in a single plot. In additional to these nests, there is another small nest measuring 2 m in radius, which is used to count saplings (i.e., trees measuring <10 cm in diameter at breast height (dbh) and >1.3 m in height). Clustering of plots at each sampling unit was recommended for natural forest areas and areas that have been selectively logged. The sampling system was designed in such a way to make the data collection process easier and faster, but reliable and representative for a particular forest stratum.

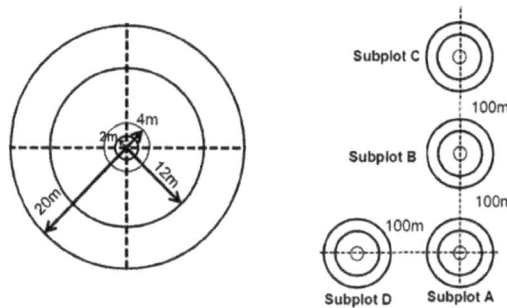

**Figure 3.** Layout of the sampling plot and sampling design of a cluster of sample.

A forest ecosystem normally has five terrestrial carbon pools, which are; (i) aboveground living biomass, (ii) belowground living biomass, (iii) deadwood, (iv) non-tree vegetation and litters, and (v) soil. However, one of the most significant carbon pools is aboveground biomass as it the easiest and the most practical pool to assess, while being representative to an ecosystem. Aboveground biomass comprises all the living components of a tree, including stems, branches, and leaves. Allometric functions are the best way that AGB can be estimated. In this study, a published allometric function for dry inland forest in Asia region was used to estimate the AGB of living trees [32].

$$AGB = [\exp(-1.803 - 0.976E + 0.976\ln(\rho) + 2.673\ln(D) - 0.0299[\ln(D)]^2]} \tag{3}$$

AGB denotes aboveground biomass (kg/tree), *E* represents bioclimatic variable, *ρ* is wood specific gravity/wood density, and *D* is dbh.

**Table 2.** Summary of living trees measurement in a plot.

| Nest Radius (m) | Size | Diameter at Breast Height, dbh (cm) |
|---|---|---|
| 2 | Sapling | <5 cm (dbh) & >1.3 m (height) |
| 4 | Small | 10.0–19.9 |
| 12 | Medium | 20.0–39.9 |
| 20 | Large | ≥40.0 |

A total number of 332 plots have been surveyed between years 2014 and 2016 and were used as sample plots information for this study. The forest survey was conducted in a number of field trips that cover mainly the central parts of Peninsular Malaysia. The States include Terengganu, Pahang, Johor, Negeri Sembilan, Selangor, Perak, Kelantan, and Perlis in the north. In each plot, every tree which meets the dbh size in the nest radius was inventoried. Species of every stand being inventoried was also recorded. Position (coordinate) of each plot was recorded at the center by using hand-held Global Positioning System (GPS) (Trimble Inc., Sunnyvale, CA, USA). The locations of all plots were post-processed by using base position data from the Department of Survey and Mapping Malaysia to ensure the accuracy of the position acquired. Locations of the sample plots are shown in Figure 4 and a summary of the sample plots is given in Table 3.

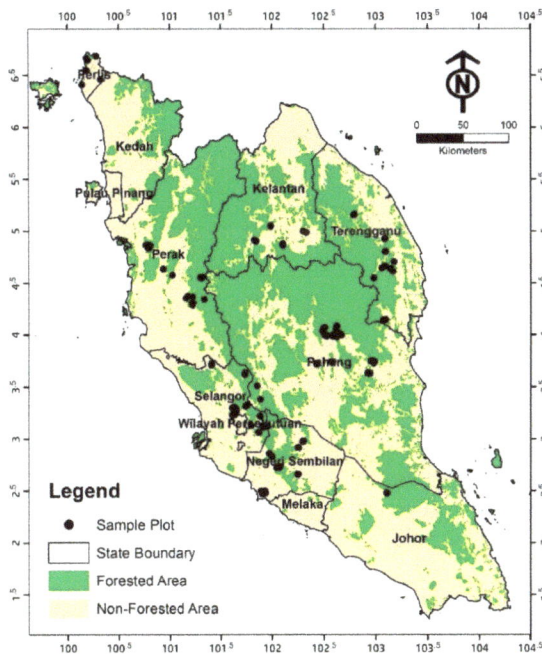

**Figure 4.** Distribution of ground sample plots within the study area.

**Table 3.** Summary of distribution of sample plots within the study area.

| State | No. of Plot | | | Total |
|---|---|---|---|---|
| | Lowland | Hill | Upper Hill | |
| Perlis | 8 | 2 | 0 | 10 |
| Terengganu | 48 | 18 | 4 | 70 |
| Pahang | 46 | 18 | 8 | 72 |
| Johor | 6 | 0 | 0 | 6 |
| Negeri Sembilan | 40 | 14 | 0 | 54 |
| Selangor | 42 | 18 | 8 | 68 |
| Perak | 20 | 8 | 4 | 32 |
| Kelantan | 12 | 8 | - | 20 |
| Total | 222 | 86 | 24 | 332 |

## 2.5. Correlation Analysis

The backscatter values from both PALSAR-2 and Sentinel-1A were extracted from the images, which represented HH, HV, VV, and VH polarizations. The AGB at the sample plots on the ground was then correlated with the corresponding backscatter of these polarizations by using linear regression method. Instead of using the single polarization as a variable, several other variables have been derived by manipulating these single polarizations. This manipulation was performed to produce variety of image variables and that to examine the roles of polarization in estimating AGB. Table 4 lists the variables that have been derived from the individual PALSAR-2 and Sentinel-1A and combination of polarizations from both sensors. These variables act as predictors to the AGB at the sample plots.

**Table 4.** Variables that were derived from PALSAR-2 and Sentinel-1A.

| Variable | PALSAR-2 | Sentinel-1A |
|---|---|---|
| Single polarization | HH<br>HV | VV<br>VH |
| Polarization multiplicative | HH × HV | VV×VH |
| Simple polarization ratio | HH − HV<br>HV − HH | VV − VH<br>VH − VV |
| Polarization averaging | (HH + HV)/2 | (VV + VH)/2 |
| Square root of multiplicative | $(HH \times HV)^{1/2}$ | $(VV \times VH)^{1/2}$ |
| Combination of polarizations | HH − VV<br>HV − VH<br>(HH + HV)/(VV + VH)<br>(VV − VH)/(VH − HV)<br>(HH + HV + VV + VH)/4<br>(HH × HV)/(VV × VH)<br>$(HH \times HV \times VV \times VH)^{1/4}$ | |

All polarizations are in sigma nought ($\sigma^0$, dB).

Simple linear regression method was used to investigate the relationship between the AGB and the image variables. Multiple linear regressions were also applied to the variables to observe whether the combination of polarizations from both PALSAR-2 and Sentinel-1A able to improve strength of the correlations. Manipulation of polarization of individual PALSAR-2 and Sentinel-1A as well as combination of both sensors were tested and multiple variable equations have been produced. The relationship between backscatter and AGB is represented in a common linear function as $y = ax + b$, $x$ and $y$ denote image variables and AGB, respectively and $a$ and $b$ are the equation coefficients. The strength of the relationship was measured by the derived coefficient of correlation ($R^2$). The greater $R^2$ indicates a stronger relationship between two variables; the value $R^2$ of 0 means no correlation and 1 is a perfect correlation. In this case, the prediction equation with the highest $R^2$ was selected to estimate AGB within the entire study area.

Studies [9,11,33] have demonstrated that PALSAR polarization data actually has a logarithmic relationship with AGB. Therefore, instead of employing linear regression only, the study also attempted to correlate the AGB with the polarizations in a non-linear form. However, this method was applied only on the individual polarizations, i.e., HH, HV, VV, and VH. Similar to simple linear regression, the estimation models used AGB as independent variable to observe the sensitivity of the backscatter to the AGB. The relationship between backscatter and AGB is commonly represented in exponential an function as $y = a \times e^{(xb)}$, where $x$ and $y$ denote image variables and AGB, respectively and $a$ and $b$ are the equation coefficients.

## 2.6. Validation Approach

The study used K-fold cross validation method to evaluate the performance of the best prediction model derived from PALSAR-2, Sentinel-1A, and combination of both data. This method provided better indication on the prediction performance than the common residual method. Residual method does not provide an indication as to how well the model makes new predictions over new sample data, but this method does. In this study, 10-fold cross validation method [34] was used where all sample plots data were randomly grouped into 10 groups. One group was used as a testing set while the other nine groups were used in developing the model. The root mean square error (RMSE) was calculated using the testing set. This process was iterated 10 times where each group was used as a testing set once. Then, the average of all RMSEs was calculated to get the overall RMSE of that model.

## 3. Results and Discussion

### 3.1. Satellite Datasets

The images that were used for analysis have been calibrated, geometrically and radiometrically corrected, and topographically normalized. Examples of the images that went through all the pre-processing are depicted in Figure 5. The topographic normalization outcome is also shown in Figure 6. The study found that these processes are necessary and must be done on any SAR images before further analyses are carried out.

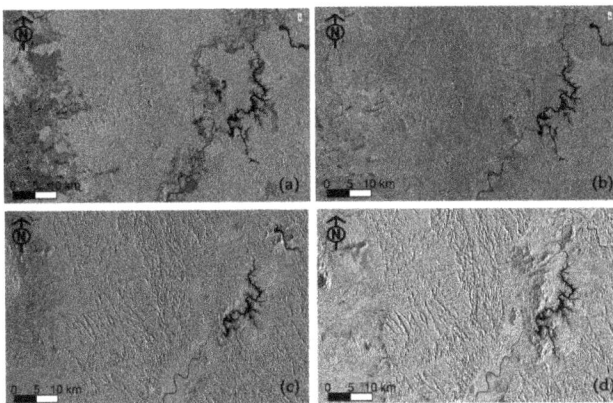

**Figure 5.** Corrected images for all polarizations of (**a**) PALSAR-2 HV; (**b**) PALSAR-2 HH; (**c**) Sentinel-1A VH; and (**d**) Sentinel-1A VV, displayed in backscatter values.

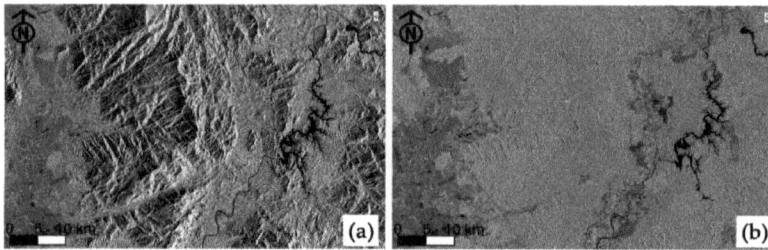

**Figure 6.** Topographic effect on SAR images (**a**), which has been normalized (**b**).

### 3.2. Forest-Non-Forest Classification

The classification that was carried out over the HH and HV polarizations and all the manipulations found that PALSAR-2 images have different capability in defining forests. The study demonstrated that the most effective polarization for FNF classification was the HV. However, the HH polarization was found effective on delineating plantation areas, such as rubber and teak, because the orientation of the plantations is systematic and homogenous, which can be interpreted well by the HH polarization. The classification was made based on the major forest types found in Peninsular Malaysia and the results are summarized in Table 5. However, this study concentrated only in lowland, hill, and upper hill dipterocarp forests. The breakdowns of these forest types are summarized in Table 6. The classification results were compared with the land use map for the year 2014 that was produced by the Department Agriculture Peninsular Malaysia and the classification accuracy was attained at 91.3% with a kappa coefficient of 0.88. The remaining 8.7% belonged to errors due to misclassification of secondary forest and rubber plantation as defined on the land use map. The results were reliable because the classification interested only in distinguishing forests from other land covers.

**Table 5.** Extents of forests in Peninsular Malaysia.

| Forest Type | Extent (ha) |
| --- | --- |
| Inland | 5,525,034 |
| Peat swamp | 264,578 |
| Mangrove | 106,198 |
| Total | 5,895,810 |

**Table 6.** Forest types within the study area.

| Forest Type | Lowland Dipterocarp | Hill Dipterocarp | Upper Hill Dipterocarp | Total (ha) |
| --- | --- | --- | --- | --- |
| Extents (ha) | 2,704,816 | 2,004,991 | 547,588 | 5,257,395 |
| Percentage (%) | 51.5 | 38.1 | 10.4 | 100 |

### 3.3. Forest Survey Results

Aboveground biomass within all the 332 sample plots have been estimated at plot level. In general, the average AGB was 399.42 Mg ha$^{-1}$ within the range between 35.57 and 615.50 Mg ha$^{-1}$ and the standard deviation of 127.82 Mg ha$^{-1}$. AGB of small trees (dbh 10–19.9 cm) contributes only about 15% of the total AGB. However, trees under this category were plenty in terms of number. Figure 7 shows the relationship between the number of trees and AGB in a hectare of the forest. The AGB is actually stored in the huge trees measuring dbh from 40 cm and above. Although the number of huge trees is low, the amount of AGB within these trees is large.

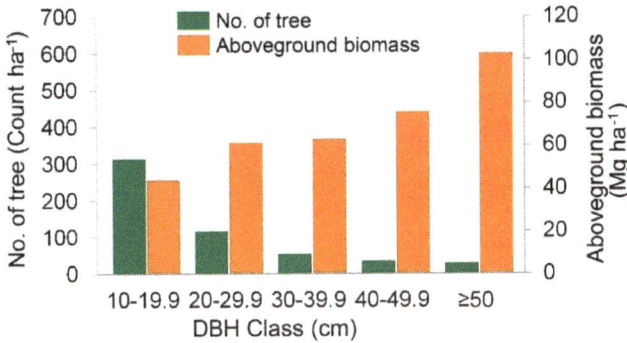

**Figure 7.** Relationship between tree size, number of trees, and AGB in a hectare of dipterocarp forest.

*3.4. Correlation Analysis*

Backscatter values from all polarizations have been extracted at all sample plots and the distribution is depicted in Figure 8. The boxplot indicates that PALSAR-2 basically had stronger backscatter over the sample plots at both polarizations as compared to Sentinel-1A polarizations. Higher variation of PALSAR-2 HV polarization indicates the capability in discriminating AGB level. On the other hand, for Sentinel-1 data, VV polarization is more sensitive to the forest as compared to VH. These backscatter values were used in the correlation analysis.

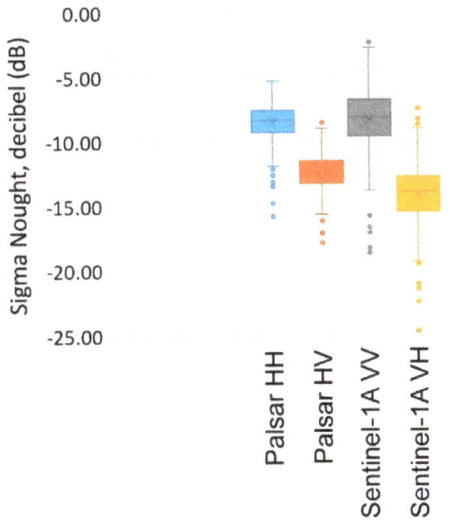

**Figure 8.** Backscatter values from all polarizations at sample plots of PALSAR-2 (HH and HV) and Sentinel-1 (VV and VH).

3.4.1. Simple Linear Regression

The results indicated that all variables showed weak linear relationships with AGB even though the correlations were significant at 95% confidence level. Table 7 summarized the correlation strength of all variables derived from the polarizations of PALSAR-2 and Sentinel-1A. The corresponding scatter plots listed in the table are shown in Figure 9. The results proved that the manipulation of polarizations from a single sensor slightly improve the correlation strength. Further improvement was attained when the polarizations are combined together, from one either single sensor or integration of both sensors.

**Table 7.** Summary of the AGB prediction equations produced from simple linear regression.

| Sensor | Variable | Scatter Plot | Prediction Equation | $R^2$ |
|---|---|---|---|---|
| PALSAR-2 | HH | a | 28.59HH + 641.68 | 0.119 |
| | HV | b | 47.21HV + 978.87 | 0.276 |
| | HH × HV | c | −1.98(HH × HV) + 608.64 | 0.223 |
| | HH − HV | d | −6.5969(HH − HV) + 369.08 | 0.005 |
| | HV − HH | e | 6.5969(HV − HH) + 369.08 | 0.005 |
| | (HH + HV)/2 | f | 43.49((HH + HV)/2) + 850.56 | 0.219 |
| | (HH × HV)$^{1/2}$ | g | −40.72((HH × HV)$^{1/2}$) + 813.88 | 0.201 |
| Sentinel-1A | VV | h | 15.56VV + 527.84 | 0.091 |
| | VH | i | 8.48VH + 518.33 | 0.023 |
| | VV × VH | j | −0.69(VV × VH) + 482.13 | 0.090 |
| | VV − VH | k | 13.78(VV − VH) + 319.88 | 0.041 |
| | VH − VV | l | −13.328(VH − VV) + 321.76 | 0.038 |
| | (VV + VH)/2 | m | 14.13((VV + VH)/2) + 556.84 | 0.062 |
| | (VV × VH)$^{1/2}$ | n | −14.46((VV × VH)$^{1/2}$) + 553.95 | 0.073 |
| Combination | HH − VV | o | −4.11(HH − VV) + 398.50 | 0.004 |
| | HV − VH | p | 6.48(HV − VH) + 388.08 | 0.015 |
| | (HH + HV)/(VV + VH) | q | −31.53((HH + HV)/(VV + VH)) + 429.95 | 0.0001 |
| | (VV − VH)/(VH − HV) | r | −590.85((VV − VH)/(VH − HV)) + 269.13 | 0.099 |
| | (HH + HV + VV + VH)/4 | s | 36.98((HH + HV + VV + VH)/4) + 797.17 | 0.176 |
| | (HH × HV)/(VV × VH) | t | −0.91((HH × HV)/(VV × VH)) + 400.39 | 0.003 |
| | (HH × HV × VV × VH)$^{1/4}$ | u | −35.30((HH × HV × VV × VH)$^{1/4}$) + 764.90 | 0.177 |

All polarizations are in sigma nought ($\sigma^0$, dB). All correlations are significant at $p < 0.05$.

### 3.4.2. Multiple Linear Regression

Synergy of the prediction has been obtained when the variables were integrated into an empirical prediction equation derived from multiple line regression. This method was applied to the single PALSAR-2, Sentinel-1A polarization, and also to the variables from the combination of both PALSAR-2 and Sentinel-1A. The best three models have been produced as summarized in Table 8. Evidently the combination of PALSAR-2 L-band and Sentinel-1A able to strengthen the relationship between AGB and the polarization, thus improving the accuracy of estimates. Both datasets have complemented to each other that eliminated the effects of backscattering diffusion.

**Table 8.** The best correlations derived from multiple regression from a single sensor and combination of sensors.

| Sensor | Prediction Equation | $R^2$ |
|---|---|---|
| PALSAR-2 | 146.90HH + 169.78HV − 7.03(HH × HV) + 416.96(HH × HV)$^{1/2}$ + 227.07 | 0.342 |
| Sentinel-1A | −17.040VH − 2.344(VV × VH) + 24.327(HH × HV)$^{1/2}$ + 181.918 | 0.138 |
| Combination | −10.877VH − 13.292(HH × HV)$^{1/2}$ + 139.702HH + 162.287HV − 6.526(HH × HV) + 394.502(HH × HV)$^{1/2}$ + 238.524 | 0.356 |

All polarizations are in sigma nought ($\sigma^0$, dB). All correlations are significant at $p < 0.05$.

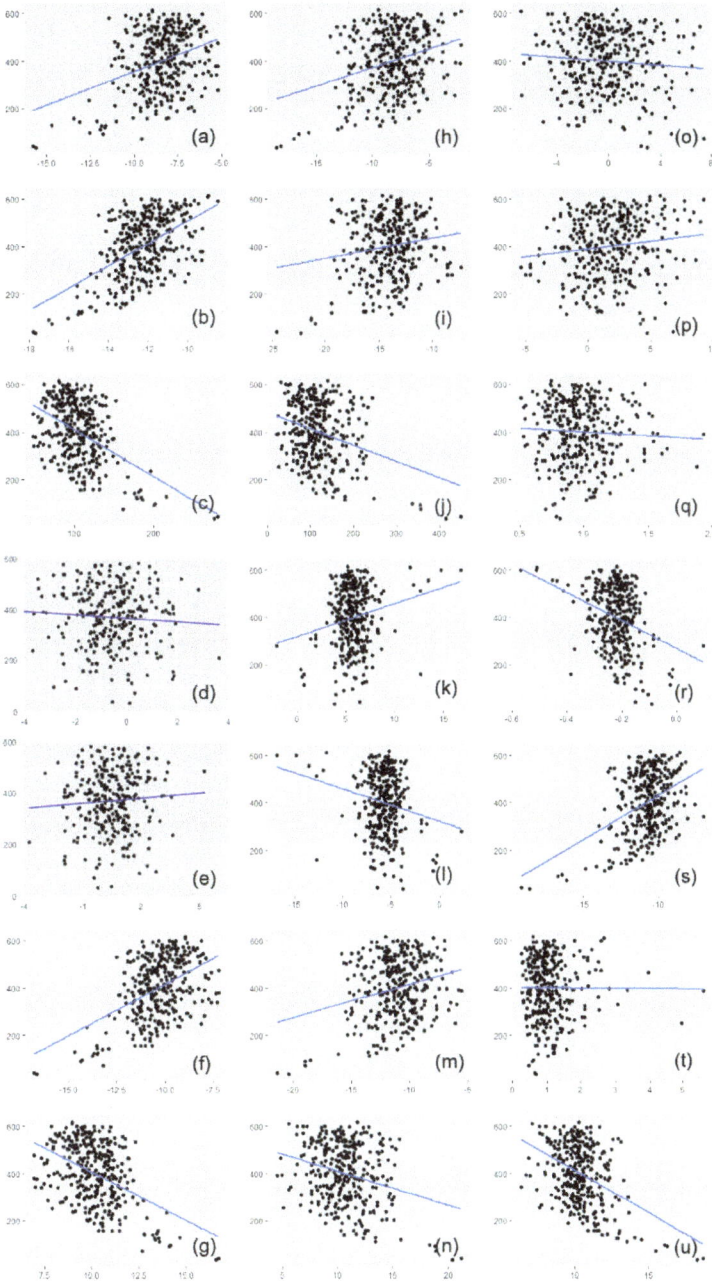

**Figure 9.** Scatter plots of simple linear correlations between AGB (*y*-axis) and image variables (*x*-axis).

3.4.3. Non-Linear Regression

Referring to the correlations listed in Table 9 and depicted in Figure 10, the backscatter of PALSAR-2 HV polarization gave better $R^2$ as compared to the HV as well as Sentinel-1A VV and VH.

The HV backscatter ranged from −1 to −20 dB and the saturation point concentrated −12 dB. Rapid increment occurred, especially at lower biomass level (i.e., up to 200 Mg ha$^{-1}$), and then decreased towards higher AGB. The trend line became almost constant when the AGB exceeded 200 Mg ha$^{-1}$. It was obvious that the estimation uncertainties are larger at AGB > 200 Mg ha$^{-1}$. The results were even worse for HH polarization.

**Table 9.** Summary of non-linear correlation between AGB and individual polarization.

| Sensor | Polarization | Prediction Equation | $R^2$ |
|--------|--------------|---------------------|-------|
| PALSAR-2 | HH | $y = 1043 \times e^{0.1215x}$ | 0.2058 |
|          | HV | $y = 3114.2 \times e^{0.173x}$ | 0.3502 |
| Sentinel-1A | VV | $y = 644.1 \times e^{0.0664x}$ | 0.1596 |
|             | VH | $y = 718.72 \times e^{0.0469x}$ | 0.0749 |

All correlations are significant at $p < 0.05$.

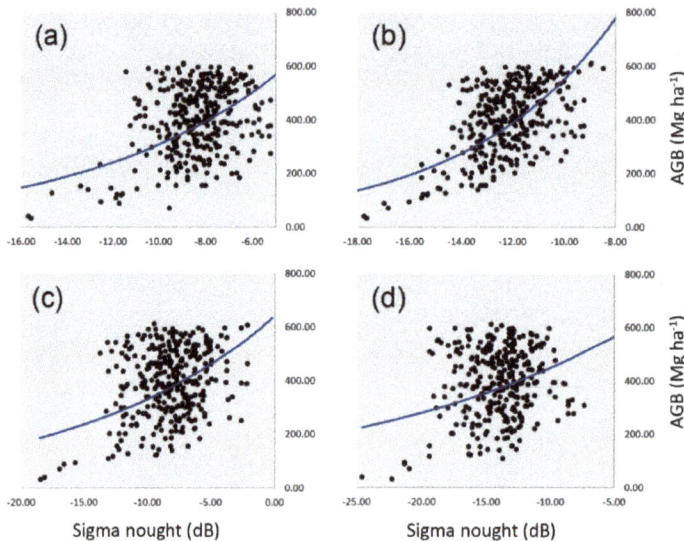

**Figure 10.** Scatter plots of non-linear correlations between AGB and single polarization of (**a**) PALSAR-2 HH; (**b**) PALSAR-2 HV; (**c**) Sentinel-1A VV; and (**d**) Sentinel-1A VH.

It has been reported that, at a given polarization and incidence angle, the saturated backscatter value for forest was within a small range of backscatter [11]. The dynamic range is determined primarily by the backscatter at low levels of AGB. It increases with decreasing frequency and it is higher at HV compared to HH polarization. Similarly, Sentiel-1A polarizations saturated quickly at AGB lower than 100 Mg ha$^{-1}$. The Sentinel-1A VV and VH backscatter ranged from −18 to −3 dB and −24 to −8 saturated at −8 dB and −14 dB, respectively. This was because the C-band wavelength is shorter than the L-band, and thus not very sensitive to the AGB at high level. Figure 11 illustrates how L- and C-bands interact with the forest canopy structure that influence the strength of the backscatter at high biomass forest.

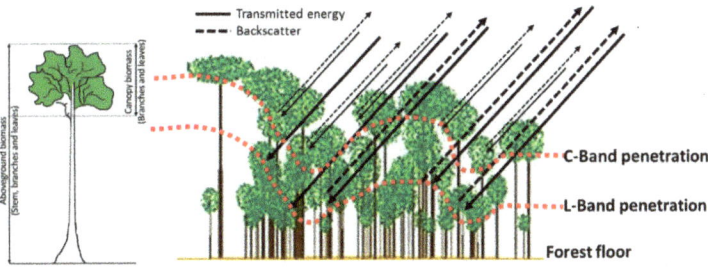

**Figure 11.** Common interaction of SAR L- and C-bands with the forest structure.

The accuracy was also mostly influenced by the tree density, soil surface roughness, soil moisture, tree sizes, and the layering effects of the SAR itself [35]. An experiment has found that the backscattering intensity interacted only with trees of dbh larger than 15 cm. These stands are actually dominating the higher canopies, which gave the best response to backscatter in HV polarization of L-band [13]. Other factors—such as orientation of the forest, polarimetry, incidence angle, and crown structure—also play important role in the estimated biomass [36,37].

### 3.4.4. The Combination Effects

Referring to Figures 8 and 11, the responses of PALSAR L-band and Sentinel-1A C-band towards AGB are different in terms of strength and variation. L-band observations penetrate more into the forest canopy and between branches and spaces, compared to the C-band, which only interacts with top canopy layers before it is scattered back or extinct. In tropical forests, volume and volume-surface scattering dominated the HH in while volumetric scattering due to dense vegetation cover dominated the HV by [38]. It is also likely that forest has a higher amount of canopy variability influencing scattering due to significant surface roughness as observed by the cross-pol (HV) term. Since structure influences the cross-pol term, forest areas that undergo selective harvesting are theoretically observable by PALSAR-2 HV, but not by Sentinel-1A, unless there is excessive timber extraction from the forest. SAR sensors can receive a relatively higher amount of surface scattering in low-density forest rather than a majority of scattering from trunks and trees or branches and crowns. Therefore, stand density, basal area, and AGB influence these relationships; although variability remains low regardless of height of the stands.

Since most of the sample plots were located in dense and mature forest (of AGB $\geq$ 200 Mg ha$^{-1}$), the variation of backscatter from both PALSAR-2 and Sentinel-1A polarizations are within the saturation threshold. Except for a number of sample plots that were located inside the secondary and logged forests, which contained relatively lower AGB than that inside the dense forest. Consequently, these factors have influenced the scatterplots in the correlations. The presence of C-band in the combination has complement the L-band at lower part of AGB (<200 Mg ha$^{-1}$) forest thus produce a better prediction overall. Taking the best linear correlations from PALSAR-2 HV and Sentinel-1A VV, with $R^2$ 0.276 and 0.091 respectively, the $R^2$ increased to 0.356 when combined. This has increased the explained variance by about an average of 17.25%. Although a single PALSAR-2 HV polarization from the non-linear correlation can predict the AGB with an $R^2$ of 0.3502, the multiple linear correlations remain stronger even with an increase of explained variance by about 0.58%. In addition, the combination of L-band polarizations, as well as L- and C-band fusion has proven to be successful, thus confirming the hypothesis of the study.

### 3.5. Estimated AGB and Mapping

The study demonstrated that the combination of polarizations from PALSAR-2 L-band and Sentinel-1A C-band provided advantages for AGB estimation in dipterocarp forest with relatively high AGB. The equation that was derived from the multi linear regression resulted from the best combination of PALSAR-2 and Sentinel-1A polarization, which gives the highest correlation value of 0.356, was used to estimate the AGB within the entire study area. The equation is expressed as

$$\text{AGB (Mg ha}^{-1}) = -10.877\text{VH} - 13.292(\text{HH} \times \text{HV})^{1/2} + 139.702\text{HH} + \\ 162.287\text{HV} - 6.526(\text{HH} \times \text{HV}) + 394.502(\text{HH} \times \text{HV})^{1/2} + 238.524 \tag{4}$$

By using this equation, AGB within the entire study area has been retrieved and mapped. Figure 12 shows the spatial distribution of AGB in the study area. From the map, the total AGB in about 5.25 million ha of the study area was estimated at 1,821,214,202 Mg over the year 2016. Figure 13 summarizes the distribution of AGB in the study area, represented by histogram of frequency of pixel occurrences. The distribution was found to be normal throughout the entire study area. Further classification was made to the AGB distribution, reported in intervals as shown in Figure 14. More than half of the study area comprised AGB within the range of 300–400 Mg ha$^{-1}$. The highest AGB was concentrated mainly in the northern part of Pahang and southern part of Kelantan, where the largest National Park in Peninsular Malaysia is located. High density of AGB occurred also in the northern part of Perak where Royal Belum State Park is located. These forests are virgin and have existed for millions years and are still intact now. Variations are found scattered in other areas where there was a mix between natural virgin and logged over forests. Low density of AGB appeared near the edges of forest areas, which mostly interacted with other land use activities nearby.

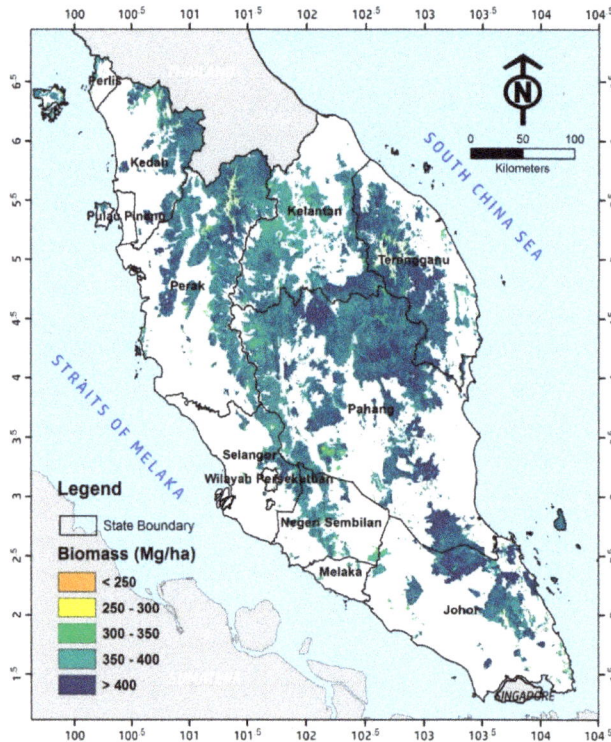

**Figure 12.** Spatially distributed map of AGB within the study area.

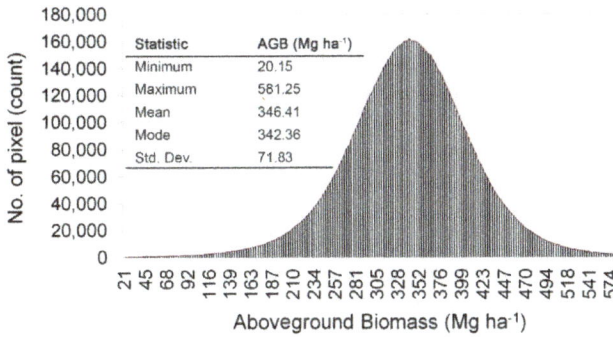

**Figure 13.** Distribution of AGB within the study area.

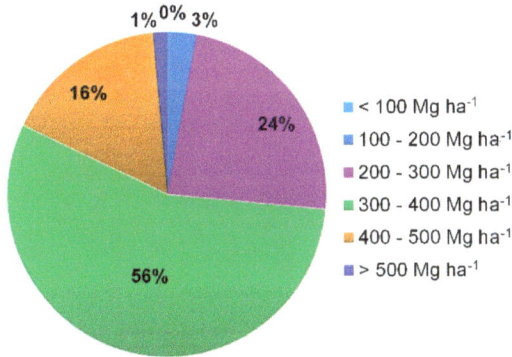

**Figure 14.** Overall breakdowns of AGB density in the study area.

*3.6. Validation of the Estimates*

Overall, the RMSE, resulted from 10-fold cross validation method, for the best model for PALSAR-2, Sentinel-1A, and combination of PALSAR-2 and Sentinel-1A are 99.10 Mg ha$^{-1}$, 111.18 Mg ha$^{-1}$, and 98.41 Mg ha$^{-1}$, respectively. The best prediction model, which was produced from the combination of polarization from both PALSAR-2 and Sentinel-1A, gave the lowest error. However, the variation of RMSE between them was considered small, which means that even if the prediction were carried out from a single PALSAR-2 or Sentinel-1A, the error will be almost at the same level. Figure 15 shows the relationship between the reference AGB calculated from ground data and predicted AGB estimated using Equation (4). The error was observed to occur at around 400 Mg ha$^{-1}$ and the distribution shows that the prediction model slightly overestimated the AGB within the study area.

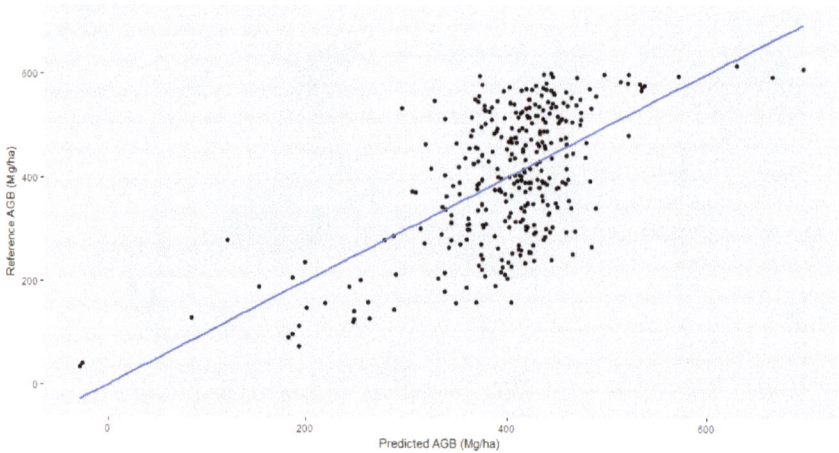

**Figure 15.** Relationship between predicted and reference AGB.

## 4. Conclusions

The study has successfully quantified the AGB over the lowland, hill, and upper hill dipterocarp forests in Peninsular Malaysia. The total AGB was estimated at about 1.82 trillion Mg over the year 2016. The extent of forested area—i.e., 5,895,810 ha—was also identified from the L-band PALSAR-2 data. The study confirmed that the synergetic of PALSAR-2 and Sentinel-1A produced better estimates than the single sensor. Although there were limitations found, the study provided an alternative for AGB retrieval that can be utilized in a practical manner to assist in the management and protection of forested areas. The study, to some extent, can also provide a significance contribution towards the MRV in the REDD+ implementation. One of greatest advantages of using the PALSAR-2 and Sentinel-1A data is the free access policy to the datasets. Free-cloud cover and rapid acquisition made them more valuable, especially for this kind of study in Malaysia.

**Acknowledgments:** This work has been undertaken within the framework of the JAXA Kyoto & Carbon Initiative (Phase 4). ALOS-2 PALSAR-2 data were provided by JAXA EORC. The financial support was to conduct the forest survey was provided by FRIM through the 11th Malaysian Plan research and development fund. Special gratitude to the State Forestry Departments of Terengganu, Pahang, Johor, Negeri Sembilan, Selangor, Perak, Kelantan, and Perlis for their supports for providing ancillary data and allowing access to some of the forest reserves for field data collection. Deepest thanks to anonymous reviewers who gave constructive comments, critically scrutinized the manuscript, and contributed to the manuscript's final state.

**Author Contributions:** H.O. led the project, contributed the main idea of this article and conducted mapping, and prepared most of the research design and findings. M.A.M. conceived the statistical analysis and led the ground survey team. A.R.K. conducted an overall assessment over the scientific merit of this study and contributed in analysis of the ground data.

**Conflicts of Interest:** The authors declare no conflict of interest.

## References

1. Intergovernmental Panel on Climate Change (IPCC). *IPCC Guidelines for National Greenhouse Gas Inventories*; Prepared by the National Greenhouse Gas Inventories Programme; Eggleston, H.S., Buendia, L., Miwa, K., Ngara, T., Tanabe, K., Eds.; Japan Institute for Global Environmental Strategies: Hayama, Japan, 2006.
2. Sessa, R.; Dolman, H. (Eds.) *Terrestrial Essential Climate Variables for Climate Change Assessment, Mitigation and Adaptation*; FAO GTOS-52; FAO: Rome, Italy, 2008.
3. Le Toan, T.; Quegan, S.; Davidson, M.W.J.; Balzter, H.; Paillou, P.; Papathanassiou, K.; Plummer, S.; Rocca, F.; Saatchi, S.; Shugart, H.; et al. The BIOMASS mission: Mapping global forest biomass to better understand the terrestrial carbon cycle. *Remote Sens. Environ.* **2011**, *115*, 2850–2860. [CrossRef]

4.  Pedro, R.V.; Wheeler, J.; Louis, V.; Tansey, K.; Balzter, H. Quantifying forest biomass carbon stocks from space. *Curr. For. Rep.* **2017**, *3*, 1–18.

5.  Gibbs, H.K.; Brown, S.; O'Niles, J.; Foley, J.A. Monitoring and estimating tropical forest carbon stocks: Making REDD a reality. *Environ. Res. Lett.* **2007**, *2*, 045023. [CrossRef]

6.  Angelsen, A.; Brown, S.; Loisel, C.; Peskett, C.; Streck, C.; Zarin, D. *Reducing Emission from Deforestation and Degradation (REDD): An Options Assessment Report*; A Report Prepared for the Government of Norway; Meridian Institute: Washington, DC, USA, 2009; p. 100.

7.  Holmgren, P. *Role of Satellite Remote Sensing in REDD*; UN-REDD Programme; MRV Working Paper 1; UN FAO: Rome, Italy, 2008.

8.  Asner, G.P. Cloud cover in Landsat observations of the Brazilian Amazon. *Int. J. Remote Sens.* **2001**, *22*, 3855–3862. [CrossRef]

9.  Lucas, R.; Armston, J.; Fairfax, R.; Fesham, R.; Accad, A.; Carreiras, J.; Kelley, J.; Bunting, P.; Clewley, D.; Bray, S.; et al. An evaluation of the ALOS PALSAR L-band backscatter—Aboveground biomass relationship Queensland, Australia: Impacts of surface moisture condition and vegetation structure. *IEEE J. Sel. Top. Appl. Earth Obs. Remote Sens.* **2010**, *3*, 576–593. [CrossRef]

10. Avtar, R.; Suzuki, R.; Takeuchi, W.; Sawada, H. PALSAR 50 m mosaic data based national level biomass estimation in Cambodia for implementation of REDD+ mechanism. *PLoS ONE* **2013**, *8*, e74807. [CrossRef] [PubMed]

11. Hamdan, O.; Mohd, H.I.; Khali Aziz, H.; Norizah, K.; Helmi Zulhaidi, M.S. Determining L-band saturation level for aboveground biomass assessment of dipterocarp forests in Peninsular Malaysia. *J. Trop. For. Sci.* **2015**, *27*, 388–399.

12. Imhoff, M.L. A theoretical analysis of the effect of forest structure on synthetic aperture radar backscatter and the remote sensing of biomass. *IEEE Trans. Geosci. Remote Sens.* **1995**, *33*, 341–352. [CrossRef]

13. Hamdan, O.; Mohd, H.I.; Khali Aziz, H. Combination of SPOT-5 and ALOS PALSAR images in estimating aboveground biomass of lowland Dipterocarp forest. In *IOP Conference Series Earth and Environmental Science*; Institute of Physics: London, UK, 2014; p. 6.

14. Chenli, W.; Zheng, N.; Xiaoping, G.; Zhixing, G.; Pifu, C. Tropical forest plantation biomass estimation using RADARSAT-SAR and TM data of South China. In Proceedings of the Fourth International Symposium on Multispectral Image Processing and Pattern Recognition (MIPPR), Wuhan, China, 31 October–2 November 2005; Liangpei, Z., Jianqing, Z., Mingsheng, L., Eds.; Volume 6043, pp. 61–69.

15. Patenaude, G.; Hill, R.A.; Milne, R.; Gaveau, D.L.A.; Briggs, B.B.J.; Dawson, T.P. Quantifying forest aboveground carbon content using lidar remote sensing. *Remote Sens. Environ.* **2004**, *93*, 368–380. [CrossRef]

16. Le Toan, T.; Quegan, S.; Woodward, I.; Lomas, M.; Delbart, N.; Picard, C. Relating radar remote sensing of biomass to modeling of forest carbon budgets. *Clim. Chang.* **2004**, *76*, 379–402. [CrossRef]

17. Proisy, C.; Mougin, E.; Fromard, F.; Karam, M.A. Interpretation of polarimetric radar signatures of mangrove forests. *Remote Sens. Environ.* **2000**, *71*, 56–66. [CrossRef]

18. Simard, M.; DeGrandi, G.; Saatchi, S.; Mayaux, P. Mapping tropical coastal vegetation using JERS-1 and ERS-1 radar data with a decision tree classifier. *Int. J. Remote Sens.* **2002**, *23*, 1461–1474. [CrossRef]

19. Lucas, R.; Moghaddam, M.; Cronin, N. Microwave scattering from mixed-species forests, Queensland, Australia. *IEEE Trans. Geosci. Remote Sens.* **2004**, *42*, 2142–2159. [CrossRef]

20. Hamdan, O.; Khali, A.H.; Abd, R.K. Remotely sensed L-Band SAR data for tropical forest biomass estimation. *J. Trop. For. Sci.* **2011**, *23*, 318–327.

21. Simard, M.; Hensley, S.; Lavalle, M.; Dubayah, R.; Pinto, N.; Hofton, M. An empirical assessment of temporal decorrelation using the uninhabited aerial vehicle synthetic aperture radar over forested landscapes. *Remote Sens.* **2012**, *4*, 975–986. [CrossRef]

22. Englhart, S.; Keuck, V.; Siegert, F. Aboveground biomass retrieval in tropical forests—The potential of combined X- and L-band SAR data use. *Remote Sens. Environ.* **2011**, *115*, 1260–1271. [CrossRef]

23. Reiche, J.; Lucas, R.; Anthea, L.; Mitchell, J.V.; Dirk, H.; Hoekman, J.H.; Kellndorfer, J.M.; Rosenqvist, A.; Eric, A.L.; Woodcock, C.E.; et al. Combining satellite data for better tropical forest monitoring. *Nat. Clim. Chang.* **2016**, *6*, 120–122. [CrossRef]

24. Morel, A.C.; Saatchi, S.; Malhi, Y.; Berry, N.J.; Banin, L.; Burslem, D.; Nilus, R.; Ong, R.C. Estimating aboveground biomass in forest and oil palm plantation in Sabah, Malaysian Borneo using ALOS PALSAR data. *For. Ecol. Manag.* **2011**, *262*, 1786–1798. [CrossRef]

25. Morel, A.C.; Fisher, J.B.; Malhi, Y. Evaluating the potential to monitor aboveground biomass in forest and oil palm in Sabah, Malaysia, for 2000–2008 with Landsat ETM+ and ALOS-PALSAR. *Int. J. Remote Sens.* **2012**, *33*, 3614–3639. [CrossRef]
26. Forestry Department Peninsular Malaysia. *Annual Report 2014*; Forestry Department Peninsular Malaysia: Kuala Lumpur, Malaysia, 2015.
27. Hoekman, D.H.; Reicheb, J. Multi-model radiometric slope correction of SAR images of complex terrain using a two-stage semi-empirical approach. *Remote Sens. Environ.* **2015**, *156*, 1–10. [CrossRef]
28. Canty, M.J. *Image Analysis, Classification and Change Detection in Remote Sensing: With Algorithms for ENVI/IDL and Python*, 3rd ed.; CRC Press: Boca Raton, FL, USA, 2014.
29. Shimada, M. Model-based Polarimetric SAR calibration method using forest and surface-scattering targets. *IEEE Trans. Geosci. Remote Sens.* **2011**, *49*, 1712–1733. [CrossRef]
30. Sarker, M.; Rahman, L.; Janet, N.; Baharin, A.; Busu, I.; Alias, A.R. Potential of texture measurements of two-date dual polarization PALSAR data for the improvement of forest biomass estimation. *ISPRS J. Photogramm. Remote Sens.* **2012**, *69*, 146–166. [CrossRef]
31. Walker, S.M.; Pearson, T.R.H.; Casarim, F.M.; Harris, N.; Petrova, S.; Grais, A.; Swails, E.; Netzer, M.; Goslee, K.M.; Brown, S. *Standard Operating Procedures for Terrestrial Carbon Measurement: Version 2014*; Winrock International: Little Rock, AR, USA, 2012.
32. Chave, J.; Maxime, R.E.; Alberto, B.; Chidumayo, E.; Matthew, S.C.; Welington, B.C.D.; Alvaro, D.; Tron, E.; Philip, M.F.; Rosa, C.G.; et al. Improved allometric models to estimate the aboveground biomass of tropical trees. *Glob. Chang. Biol.* **2014**, *20*, 3177–3190. [CrossRef] [PubMed]
33. Liesenberg, V.; Gloaguen, R. Evaluating SAR polarization modes at L-band for forest classification purposes in Eastern Amazon, Brazil. *Int. J. Appl. Earth Obs. Geoinf.* **2013**, *21*, 122–135. [CrossRef]
34. McLachlan, G.J.; Do, K.A.; Christophe, A. *Analyzing Microarray Gene Expression Data*; Wiley: Hoboken, NY, USA, 2004.
35. Quinones, M.J.; Hoekman, D.H. Exploration of factors limiting biomass estimation by polarimetric radar in tropical forests. *IEEE Trans. Geosci. Remote Sens.* **2004**, *42*, 86–104. [CrossRef]
36. Watanabe, M.; Shimada, M.; Rosenqvist, A.; Tadono, T.; Matsuoka, M.; Romshoo, S.A.; Ohta, K.; Furuta, R.; Nakamura, K.; Moriyama, T. Forest structure dependency of the relation between L-band $\sigma^0$ and biophysical parameters. *IEEE Trans. Geosci. Remote Sens.* **2006**, *44*, 3154–3165. [CrossRef]
37. Guo, Z.; Ni, W.; Sun, G. Analysis of the effect of crown structure changes on backscattering coefficient using modelling and SAR data. *IEEE Trans. Geosci. Remote Sens.* **2009**, *4*, 386–389.
38. Nathan, T.; Lindsay, L.; William, S.; Meng, Z. Regional mapping of plantation extent using multisensor imagery. *Remote Sens.* **2016**, *8*, 236.

*applied*
*sciences*

MDPI

*Article*

# Comparison of Oil Spill Classifications Using Fully and Compact Polarimetric SAR Images

Yuanzhi Zhang [1,*], Yu Li [2,*], X. San Liang [1] and Jinyeu Tsou [3]

1    School of Marine Sciences, Nanjing University of Information Science and Technology,
     Nanjing 210044, China; sanliang@courant.nyu.edu
2    School of Information and Communication Engineering, Beijing University of Technology,
     Beijing 100021, China
3    Center for Housing Innovations, Chinese University of Hong Kong, Ma Liu Shui, Hong Kong, China;
     jinyeutsou@cuhk.edu.hk
*    Correspondence: yuanzhizhang@cuhk.edu.hk (Y.Z.); yuli@bjut.edu.cn (Y.L.); Tel.: +86-10-64807833 (Y.Z.)

Academic Editor: Juan M. Lopez-Sanchez
Received: 25 December 2016; Accepted: 8 February 2017; Published: 16 February 2017

**Abstract:** In this paper, we present a comparison between several algorithms for oil spill classifications using fully and compact polarimetric SAR images. Oil spill is considered as one of the most significant sources of marine pollution. As a major difficulty of SAR-based oil spill detection algorithms is the classification between mineral and biogenic oil, we focus on quantitatively analyzing and comparing fully and compact polarimetric satellite synthetic aperture radar (SAR) modes to detect hydrocarbon slicks over the sea surface, discriminating them from weak-damping surfactants, such as biogenic slicks. The experiment was conducted on quad-pol SAR data acquired during the Norwegian oil-on-water experiment in 2011. A universal procedure was used to extract the features from quad-, dual- and compact polarimetric SAR modes to rank different polarimetric SAR modes and common supervised classifiers. Among all the dual- and compact polarimetric SAR modes, the $\pi/2$ mode has the best performance. The best supervised classifiers vary and depended on whether sufficient polarimetric information can be obtained in each polarimetric mode. We also analyzed the influence of the number of polarimetric parameters considered as inputs for the supervised classifiers, onto the detection/discrimination performance. We discovered that a feature set with four features is sufficient for most polarimetric feature-based oil spill classifications. Moreover, dimension reduction algorithms, including principle component analysis (PCA) and the local linear embedding (LLE) algorithm, were employed to learn low dimensional and distinctive information from quad-polarimetric SAR features. The performance of the new feature sets has comparable performance in oil spill classification.

**Keywords:** oil spill; SAR data; compact polarimetric mode; image classification; feature selection

---

## 1. Introduction

Oil spill is one of the most significant sources of marine pollution. In recent years, a series of accidents continually took place and threatened the marine environment. In April 2010, during the Deepwater Horizon (DWH) accident, approximately 780,000 $m^3$ of oil, methane or other fluids were released into the Gulf of Mexico. In 2011, approximately 700 barrels of crude oil were leaked into the Bohai Sea, and about 2500 barrels of mineral oil-based mud became deposited on the seabed. In December 2013, during an accident caused by a broken oil pipe, crude oil leaked into the coastal area of Qingdao, Shandong province, and covered approximately 1000 $m^2$ of the sea surface. In addition, a large proportion of oil spills are caused every year by deliberate discharges from tankers or cargos, for the reason that there are still vessels that secretly clean their tanks or engine before entering the harbor. These accidents and illegal acts cause damage to the coastal ecosystem, emphasizing the importance of detecting oil spills in their early stages.

Although remote sensing with optical sensors can be used in oil spill detection, they are unavoidably restricted by weather and light conditions. Therefore, satellite synthetic aperture radar (SAR) data from ERS-1/2 (European Remote Sensing Satellites), ENVISAT (Environmental Satellite), ALOS (Advanced Land Observing Satellite), RADARSAT-1/2 and TerraSAR-X have been widely used to detect and monitor oil spills [1–8] due to the large spatial coverage, all-weather conditions and imaging capability during day-night times [9]. In addition, airborne SAR sensors, such as Uninhabited Aerial Vehicle Synthetic Aperture Radar (UAVSAR) developed by JPL at L-band and E-SAR (developed by DLR), have proven their potential for scientific research on ocean or land [10,11].

In SAR images, the detection of oil slick on the sea surface relies on the detection/quantification of its attenuation of Bragg scattering on the sea surface. When Bragg scattering happens, the signals from different sea surface facets interfere with each other. Moreover, according to the composite sea surface model, the roughness of the sea surface can be seen as small-scale capillary waves (contributing to Bragg scattering) superimposed on large-scale gravity waves. An illustration of this model can be seen in Figure 1. The sea surface of the oil-covered region appears smoother than its surrounding area. This is because the Bragg scattering of these areas is suppressed by the presence of hydrocarbons. However, the main backscatter from the sea surface is contributed by Bragg scattering. As a result, in SAR images the oil slick-covered area can usually be detected as a very dark (low backscattered) area.

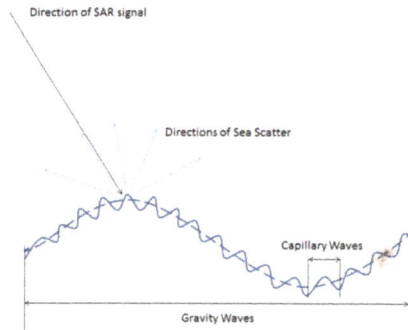

**Figure 1.** Demonstration of radar scattering from the sea surface.

In SAR images, the backscattered signal from oil spill is very similar to that from other ocean phenomena called "look-alikes" [1]. In recent years it has been demonstrated by theoretical and experimental studies the benefit of the polarimetric SAR paradigm, which explore the polarimetric SAR measurements and a proper electromagnetic modelling to distinguish light-damping surfactants from heavy-damping ones. This can be exploited as one case to sort out most of the look-alikes that are typical, such as biogenic films (slicks that are produced by marine organisms, such as fishes, algae, etc.), which normally cause very little harm to the marine environment [12,13].

The feasibility of polarimetric SAR-based oil spill classifications relies on the fact that the polarimetric mechanisms for oil-free and oil-covered sea surface are largely different [14]. Before the availability of polarimetric observations, hydrocarbons and biogenic slicks were difficult to distinguish because they damped the short gravity-capillary waves with almost the same strength [15]. However, based on different polarimetric scattering behaviors, hydrocarbons and biogenic slicks can now be better distinguished: for oil-covered areas, Bragg scattering is largely suppressed, and high polarimetric entropy can be documented. In the case of a biogenic slick, Bragg scattering is still dominant, but with a lower intensity. Thus, similar polarimetric behaviors as those of oil-free areas should be expected in the presence of biogenic films [3].

Despite being helpful to oil spill classifications, fully (or quad-) polarimetric SAR is facing the challenges of its system complexity and reduced swath width caused by the doubled pulse repetition frequency (PRF). To overcome this problem, dual polarimetric SAR systems, which transmit a single polarization signal, are often considered [16]. However, traditional dual polarimetric SAR systems transmitting only horizontal or vertical polarized signals have a limitation when acquiring the complete polarimetric behavior of selected targets.

Compared with conventional dual-polarimetric SAR modes, compact polarimetric (CP) SAR systems have higher sensitivity to the polarimetric behavior of some ground targets. Similarly, in CP SAR modes, the radar transmits only a linear combination of horizontal and vertical ($\pi/4$) or circularly ($\pi/2$, also called CTLR) polarized signals and linearly receives both horizontal and vertical polarizations. As a result, compact polarimetric SAR modes can be considered as special kinds of dual polarimetric SAR modes, and vice versa. Based on a general formalism of dual and compact polarimetric SAR data, a unified framework was proposed to analyze different CP SAR modes and its feature products [17].

Since the 2000s, CP SAR has become a new research trend [16,18,19]. In the years following the development of this technique, most studies focused on the applications of land monitoring, e.g., biomass and soil moisture estimation [20]. Recently, it began to be considered in maritime surveillance applications [21–23].

In data received via CP SAR modes, Stokes parameters and covariance matrices can be calculated from the measurement vector of SAR data, and further polarimetric analysis can be employed [24]. Some important polarimetric parameters, such as the degree of polarization (DoP), relative phase, entropy, anisotropy and $\alpha$, can also be derived [22,25,26]. It is noted that the processing method and definitions of some parameters for CP SAR data, in the process of calibration, decomposition and classification, can be different.

Some previous studies explored the possibility of taking advantage of dual- and compact polarimetric SAR data to classify oil spills and biogenic slicks [27–29]. However, there are seldom quantitative comparisons of different polarimetric SAR modes, and their performance for actual oil spill classification applications. One important benefit of Pol-SAR paradigm is its robustness, i.e., it successfully works with airborne and spaceborne SARs and for different frequencies. Due to the fine classification capability of polarimetric features, polarimetric SAR-based methods may work on a wider range of sea status (surface wind and currents). However, because of the complexity of sea surface polarimetric scattering mechanisms, it is unrealistic to consider using any single characteristic to distinguish a variety of kinds of oil spills under different conditions. As a result, a synthetic and proper use of the polarimetric characteristics is the key to the accurate detection and successful interpretation of oil slicks. Moreover, the optimum compact polarimetric SAR mode varies with the different scattering behavior of the targets and also depends on specific classification tasks. Hence, in this study, we compare the oil spill detection using quad-, dual- and compact polarimetric features using supervised oil spill classifications. The study mainly concentrated on: (a) investigating the feature selection from quad- and compact polarimetric SAR data; (b) testing the performance of these features using several supervised classification algorithms, and (c) comparing SAR data from these modes to achieve marine oil spill classifications.

## 2. Methods

### 2.1. Quad-Polarimetric SAR Mode

For quad-pol SAR data, the $2 \times 2$ scattering matrix is measured on the traditional linearly horizontal and vertical bases, which can be described by [30]:

$$S = \begin{pmatrix} S_{HH} & S_{HV} \\ S_{VH} & S_{VV} \end{pmatrix} \tag{1}$$

where the subscript $H$ and $V$ describes the transmitted and received polarization, respectively, with $H$ denoting horizontal and $V$ denoting vertical directions. For the monostatic case, the reciprocity always holds, which means that the two cross-polarized terms are identical, i.e., $S_{HV} = S_{VH}$.

The covariance matrix can be derived by:

$$
\mathbf{C} = \begin{pmatrix}
\langle S_{HH}^2 \rangle & \langle \sqrt{2} S_{HH} S_{HV}^* \rangle & \langle S_{HH} S_{VV}^* \rangle \\
\langle \sqrt{2} S_{HV} S_{HH}^* \rangle & \langle 2 S_{HV}^2 \rangle & \langle \sqrt{2} S_{HV} S_{VV}^* \rangle \\
\langle S_{VV} S_{HH}^* \rangle & \langle \sqrt{2} S_{VV} S_{HV}^* \rangle & \langle S_{VV}^2 \rangle
\end{pmatrix}
\tag{2}
$$

where * is the symbol of conjugate and "< >" stands for multilook by using an averaging window ($5 \times 5$ in this study). This $5 \times 5$ averaging window is important to obtain the statistical property of the compound target's polarization status and reduce the effect of speckle noise.

### 2.2. Feature Extraction from Quad-Polarimetric SAR Data

#### 2.2.1. Single Polarimetric Intensity

The intensity of co-polarized channel is largely used in single polarimetric SAR-based oil spill detection algorithms. In this study, $S_{VV}^2$ is considered as one of the features for its higher SNRs compared to $S_{HH}^2$ on the sea surface. The hydrocarbons on the sea surface damp the short gravity and capillary waves of the sea surface, and hence, they are usually observed as very low backscatter areas. However, very similar dark areas can also be observed from SAR images when other kinds of oil are present, such as biogenic slicks.

#### 2.2.2. $H/\alpha$ Decomposition Parameters

In 1997, Cloude and Pottier proposed a polarimetric information extraction method based on the decomposition of the $3 \times 3$ coherency matrix (3) of the target [31]:

$$
\mathbf{T} = \mathbf{U}_3 \begin{bmatrix} \lambda_1 & & \\ & \lambda_2 & \\ & & \lambda_3 \end{bmatrix} \mathbf{U}_3^H
\tag{3}
$$

where $H$ stands for transpose conjugate, and $\mathbf{U}_3$ can be parameterized by Equation (4):

$$
\mathbf{U}_3 = \begin{bmatrix}
\cos(\alpha_1) e^{j\phi_1} & \cos(\alpha_2) e^{j\phi_2} & \cos(\alpha_3) e^{j\phi_3} \\
\cos(\alpha_1)\cos(\beta_1) e^{j\delta_1} & \sin(\alpha_2)\cos(\beta_2) e^{j\delta_2} & \sin(\alpha_3)\cos(\beta_3) e^{j\delta_3} \\
\sin(\alpha_1)\sin(\beta_1) e^{j\gamma_1} & \sin(\alpha_2)\sin(\beta_2) e^{j\gamma_2} & \sin(\alpha_3)\cos(\beta_3) e^{j\gamma_3}
\end{bmatrix}
\tag{4}
$$

The three eigenvalues of the coherency matrix $\mathbf{T}$ are real numbers, arranged as $\lambda_1 > \lambda_2 > \lambda_3$, $\mathbf{U}_3$ is the unitary matrix, whose column vectors $\vec{u}_1$, $\vec{u}_2$ and $\vec{u}_3$ are the eigenvectors of $\mathbf{T}$:

$$
\mathbf{T} = \sum_{i-1}^{3} \lambda_1 \vec{u}_1 \cdot \vec{u}_1^H + \lambda_2 \vec{u}_2 \cdot \vec{u}_2^H + \lambda_3 \vec{u}_3 \cdot \vec{u}_3^H
\tag{5}
$$

The probability of three eigenvectors can be calculated by:

$$
P_i = \frac{\lambda_i}{\sum\limits_{j=1}^{3} \lambda_j}
\tag{6}
$$

The polarimetric entropy, which describes the randomness of the scattering mechanisms, can be defined as:

$$H = -\sum_{i=1}^{3} P_i \log_3(P_i) \tag{7}$$

The mean scattering angle $\alpha$ is defined by:

$$\alpha = P_1\alpha_1 + P_2\alpha_2 + P_3\alpha_3 \tag{8}$$

The entropy $H$ is a measure of the randomness of the scatter mechanism. It is base-invariant and closely related to eigenvalue $\lambda$, which represents different components of the total scatter power. For a clean sea surface, Bragg scattering dominates, so $H$ is close to 0. In contrast, for oil slick-covered areas, the scattering mechanism becomes more complex; stronger random scattering results in higher entropy values. Moreover, for biogenic slicks, although the scattering power is lower, the main scattering mechanism is still Bragg, resulting in lower entropy compared to oil-covered areas. This way, $H$ can be used to distinguish oil slicks and weak damping look-alikes.

Usually jointly used with $H$, the mean scattering angle $\alpha$ reflects the main scattering mechanism of the observed target. On slick-free sea surfaces, $\alpha$ is expected to be less than $45°$ as the Bragg scattering is dominant. In slick-covered regions, larger $\alpha$ can be measured, as a more complex scattering mechanism is present.

### 2.2.3. Degree of Polarization

Degree of polarization (DoP) is considered to be a very important parameter characterizing partially polarized EM waves. It can be derived from the Stokes vectors of any coherent radar modes, e.g., dual-pol, hybrid/compact and, of course, fully polarimetric SAR modes [32]:

$$P = \frac{\sqrt{g_{i1}^2 + g_{i2}^2 + g_{i3}^2}}{g_{i0}^2} = \left(1 - 4\frac{|\Gamma_i|}{(tr\Gamma_i)^2}\right)^{\frac{1}{2}} \tag{9}$$

where $g_i$ is Stokes vectors that can be used to describe both complete and partially polarized wave, and $i$ stands for different polarization of transmission.

$$g = \begin{bmatrix} g_0 \\ g_1 \\ g_2 \\ g_3 \end{bmatrix} = \begin{bmatrix} \langle |E_v|^2 + |E_h|^2 \rangle \\ \langle |E_v|^2 - |E_h|^2 \rangle \\ 2\mathrm{Re}\langle E_h E_v^* \rangle \\ 2\mathrm{Im}\langle E_h E_v^* \rangle \end{bmatrix} \tag{10}$$

where $E_v$ and $E_h$ is vertically and horizontally received backscatter, respectively, and $<>$ also stands for multilook by using an averaging window.

DoP measures to what extent the scattered wave is deterministic and can be described by a polarimetric ellipse with fixed parameters. On the Poincare sphere, it represents the distance between the last three components normalized Stokes vectors and the origin [32]. It is 1 for complete polarized waves and 0 for fully unpolarized waves. For clean sea surfaces and weak-damping areas, the scattering mechanism can be described by the Bragg scattering; as a result, the DoP is large. For hydrocarbon slicks, random scattering mechanisms are dominating, and much lower DoP are observed.

### 2.2.4. Ellipticity $\chi$

Ellipticity $\chi$ describes the polarization status of the scattered EM wave. From the Stokes vector, it can be calculated by:

$$\sin(2\chi) = -\frac{s_3}{ms_0} \tag{11}$$

where $m$ stands for the degree of polarization of the received EM wave.

The parameter $\chi$ can be employed as an indicator of the scattering mechanism. For slick-free sea surfaces where Bragg scattering is dominant, the sign of $\chi$ is negative. For oil-covered sea surfaces, as a more random scattering mechanism is present, $\chi$ will be larger and can become positive [28]. This feature makes $\chi$ a logical binary descriptor of slick-free vs. oil-covered areas.

### 2.2.5. Pedestal Height

Normalized radar cross-section (NRCS) $\sigma^0$ measures how detectable an object is per unit area on the ground. In the co-polarized signature of the scene, the $\sigma^0$ is a function of both the tilting angle $\Phi$ and the ellipticity $\chi$ of the polarization ellipse. The pedestal height (PH) is defined as the lowest value of all the $\sigma^0$, plotted in the co-polarized signature. The PH describes the unpolarized energy of the total scattering power and behaves as a pedestal on which the co-polarized signature is set [14,33]. The normalized pedestal height ($NPH$) can be approximately calculated as the minimum eigenvalue divided by the maximum one:

$$NPH = \frac{\min(\lambda_1, \lambda_2, \lambda_3)}{\max(\lambda_1, \lambda_2, \lambda_3)} \tag{12}$$

For clean sea surfaces, the scattering mechanism is pure Bragg, so an $NPH$ value close to 0 is expected. For an oil-covered area, however, much higher $NPH$ can be expected due to the non-Bragg scattering that reflects a more diverted scattering mechanism.

### 2.2.6. Co-Polarized Phase Difference

The co-polarized phase difference (CPD) is defined as the phase difference between the $HH$ and $VV$ channels [3]:

$$\varphi_c = \varphi_{HH} - \varphi_{VV} \tag{13}$$

From multilook SAR data, it can be also derived as:

$$\varphi_c = \arg(\langle S_{HH}S_{VV}^* \rangle) \tag{14}$$

where arg(*) stands for phase calculation.

The standard deviation of CPD has been proposed as a very efficient parameter indicating sea surface scattering mechanisms [3]. It can be estimated from $\varphi_c$ using a sliding window. On slick-free sea surfaces, the $HH$-$VV$ correlation is high, and a narrow CPD distribution is expected. This resulting CPD will have a small standard deviation, similarly for weak-damping surfactant-covered areas. In oil slicks where the Bragg scattering is weakened and other scattering mechanisms increase, the $HH$-$VV$ correlation largely decrease. As a result, the CPD pdf becomes broader, and its standard deviation largely increases.

### 2.2.7. Conformity Coefficient

The conformity coefficient $\mu$ was firstly used in compact polarimetric SAR applications for soil moisture estimations (Freeman et al., 2008). In a fully polarimetric scheme, it can be approximated as [6]:

$$\mu \cong \frac{2(\mathrm{Re}(S_{HH}S_{VV}^*) - |S_{HV}|^2)}{|S_{HH}|^2 + 2|S_{HV}|^2 + |S_{VV}|^2} \tag{15}$$

The conformity coefficient $\mu$ evaluates whether surface scattering is the dominant among all the scattering mechanisms. For a slick-free sea surface, Bragg scattering results in a very small cross-pol power and high $HH$-$VV$ correlations and $\mathrm{Re}(S_{HH}S_{VV}^*) > |S_{HV}|^2$; hence, $\mu$ is positive. However, for hydrocarbon-covered areas, as non-Bragg scattering exists, $HH$-$VV$ correlation is lower, and cross-pol component largely increases, which is very likely to have $\mathrm{Re}(S_{HH}S_{VV}^*) < |S_{HV}|^2$; hence, $\mu$ is negative. For weak-damping cases, such as biogenic slicks, since Bragg scattering is still dominant, $\text{Re}(S_{HH}S_{VV}^*) > |S_{HV}|^2$ is still valid and results in positive $\mu$. Under this rationale, conformity coefficients can be used to effectively distinguish hydrocarbon slicks from biogenic slicks.

### 2.2.8. Correlation and Coherence Coefficients

The correlation and coherence coefficients that are derived from the coherence matrix are also used for oil slick discrimination [34].

$$\rho_{HH/VV} = \left| \frac{\langle S_{HH}S_{VV}^* \rangle}{\langle S_{HH}^2 \rangle \langle S_{VV}^2 \rangle} \right| \qquad (16)$$

$$Coh = \frac{|\langle T_{12} \rangle|}{\sqrt{\langle T_{11} \rangle \langle T_{22} \rangle}} \qquad (17)$$

where $T_{ij}$ are elements of the coherence matrix **T**.

These two parameters both lie between 0 and 1. For a slick-free area, where Bragg scattering is dominant, *HH* and *VV* channels are highly correlatable, so they are expected to be very close to 1. For an oil-covered sea surface, a much lower *HH*/*VV* correlation is expected, so both the correlation and coherence coefficients are much lower.

The polarimetric SAR features above and their relative behaviors in the presence of different ocean surface targets are summarized in Table 1.

**Table 1.** Behaviors of main polarimetric SAR features on different types of surfaces. DoP, degree of polarization; CPD, co-polarized phase difference.

| Pol-SAR Features | Clean Sea Surface | Mineral Oil (Strong Damping) | Biogenic Slicks (Weak Damping) |
|---|---|---|---|
| Entropy (H) | Lower | High | Low |
| Alpha ($\alpha$) | Lower | High | Low |
| DoP | High | Low | High |
| Ellipticity | Negative | Positive | Negative |
| Pedestal Height (PH) | Lower | High | Low |
| Std. CPD | Lower | High | Low |
| Conformity Coefficient | Positive | Negative | Positive |
| Correlation Coefficient | Higher | Low | High |
| Coherence Coefficient | Higher | Low | High |
| $S_{VV}^2$ | High | Low | Low |

Note: "lower" and "higher" means that the property of the feature on a certain type of surface is close to the other surface that has the property of "low" or "high", but slightly lower or higher. "Std. CPD" stands for the standard deviation of CPD.

### 2.3. Dual- and Compact Polarimetric SAR Modes

Compact polarimetric SAR modes were proposed to solve the contradiction between polarimetric observation capabilities and the swath width, system complexity, power budget and data rate of the radar system. Actually, the idea of transmitting one polarized signal and coherently recording the backscattered signal in *H* and *V* polarimetric channels was considered by U.S. scientists as far back as 1960. In the 2000s, this operation mode was reconsidered by Souyris et al. [16] and was given the new name of "compact polarimetric" to differentiate from "fully polarimetric" or "quad-polarimetric".

Dual polarimetric (DP) SAR systems transmit a horizontal (*H*) or a vertical (*V*) linearly-polarized signal and coherently record both horizontal and vertical polarized backscattered signals. They can be treated as a special kind of compact polarimetric SAR mode. In real applications, usually *HH*/*HV* or *HV*/*VV* dual polarization modes are used, for the reason that in these modes, only the *H* or *V* polarized signal is transmitted. However, on the sea surface, the backscatter of the cross-polarized channel (*HV*) is usually much lower than the co-polarized channels [34], sometimes close to the noise floor of the radar instruments. As a result, *HH/HV* dual polarimetric modes have limited performance on oil spill classification applications. Although there is no *HH-VV* dual polarimetric SAR operating, except a special experimental imaging mode of TerraSAR-X, this mode is considered for comparative analysis in this paper.

The 2D measurements vector $\vec{K}$ of *HH/VV* dual-polarized, $\pi/2$ and $\pi/4$ compact polarimetric SAR modes are provided in Equations (18)–(20), respectively:

$$\vec{K}_{HHVV} = \begin{pmatrix} S_{HH} \\ S_{VV} \end{pmatrix} \tag{18}$$

$$\vec{K}_{pi/2} = \frac{1}{\sqrt{2}} \begin{pmatrix} S_{HH} - jS_{HV} \\ S_{HV} - jS_{VV} \end{pmatrix} \tag{19}$$

$$\vec{K}_{pi/4} = \frac{1}{\sqrt{2}} \begin{pmatrix} S_{HH} + S_{HV} \\ S_{HV} + S_{VV} \end{pmatrix} \tag{20}$$

Table 2 lists several main polarimetric SAR modes, which can be differentiated by their different transmission and receiving polarimetric combinations.

The covariance matrix can also be used to reflect the second order statistics of the dual and compact polarimetric SAR data, which can be derived from their scattering matrix by:

$$C_{CP} = 2 \left\langle \vec{K}_{CP} \vec{K}_{CP}^* \right\rangle \tag{21}$$

where $\vec{K}_{CP}$ stands for measurements vector $\vec{K}$ of different dual- and compact polarimetric SAR modes.

**Table 2.** Different polarimetric SAR modes.

| Receive / Transmit | H | V | H and V (Incoherently) | H and V (Coherently) | R and L (Coherently) |
|---|---|---|---|---|---|
| H | Single | Single | Alternating | Dual Pol | — |
| V | Single | Single | Alternating | Dual Pol | — |
| H and V | — | — | Alternating | — | — |
| 45° | — | — | — | $\pi/4$ Compact | — |
| R/L | — | — | — | $\pi/2$ Compact (Hybrid) | Dual-circular (DCP) |

Note: Blank means that at the present stage, there is not an operational SAR system with such a transmit/receive combination. R and L stand for right and left circular polarization, respectively.

### 2.4. Universal Feature Extraction from Dual- and Compact Polarimetric SAR Data

In order to explore polarimetric information, the following methods can be used to universally extract features from the measurements vectors of dual- and compact polarimetric SAR data. The features extracted from dual- and compact polarimetric modes shares similar characteristics as those derived from fully polarimetric mode, in the presence of a clean sea surface, hydrocarbons and biogenic films. Of course, some differences can also be observed between them for the reason that they carry different parts of the information of quad-pol SAR data. In the following part of this paper, they are compared and analyzed.

### 2.4.1. Elements in Measurement Vector $\vec{K}$

The elements of the measurement vector $\vec{K}$ of dual and compact polarimetric SAR modes can be derived from Equations (18)–(20):

$$\vec{K} = \begin{pmatrix} E_H & E_V \end{pmatrix}^T \tag{22}$$

where $T$ stands for the transpose.

Since for the sea surface, usually $S_{HV}^2$ is much smaller compared with the backscatter of co-polarized channels, $E_V^2$ represents close physical meaning to $S_{VV}^2$. It is selected as one of the features in classification experiments based on compact polarimetric SAR modes.

### 2.4.2. $H/\alpha$ Decomposition Parameters

Polarimetric entropy of CP SAR data can be directly calculated from the eigenvalues of the covariance matrix $C_{CP}$:

$$H = \sum_{i=1}^{2} -P_i \log_2 P_i \tag{23}$$

$$P_i = \frac{\lambda_i}{\sum_j \lambda_i} \tag{24}$$

Additionally, $\lambda_i$ ($i = 1, 2$) is the eigenvalue of coherency matrix $C_{CP}$. Entropy that is derived directly from CP SAR data has similar performance as that derived from quad-pol SAR data, in describing the complexity of the physical scattering mechanisms of targets.

Then, the mean scattering angle in CP SAR modes can be approximated by:

$$\alpha = P_1 \alpha_1 + P_2 \alpha_2 \tag{25}$$

where $\alpha_i$ can be derived from the eigenvector of the covariance of CP SAR data, similarly as in Section 2.2.

### 2.4.3. Degree of Polarization and Ellipticity

The degree of polarization and ellipticity can be similarly calculated from the Stokes vector of CP SAR mode, as introduced in Section 2.2.

### 2.4.4. Pedestal Height

Similarly, as in Section 2.2.5, pedestal height can be estimated from the eigenvalues of the covariance matrix of compact polarimetric SAR data by:

$$NPH = \frac{\min(\lambda_1, \lambda_2)}{\max(\lambda_1, \lambda_2)} \tag{26}$$

### 2.4.5. Co-Polarized Phase Difference

CPD can be proximately estimated from covariance matrix of CP SAR data by:

$$\varphi_{c(CP)} = \arg\{-iE_H E_V^*\} \tag{27}$$

Then, its standard deviation within a certain spatial window can be computed. In this paper, a window of $5 \times 5$ is applied.

### 2.4.6. Conformity Coefficient

Only for $\pi/2$ mode, the conformity coefficient is expressed as [6]:

$$Conf \cong \frac{2\mathrm{Im}(\langle E_H E_V^* \rangle)}{\langle E_H E_H^* \rangle + \langle E_V E_V^* \rangle} \tag{28}$$

### 2.4.7. Correlation Coefficient and Coherence Coefficient

Following the same rationale as in Section 2.2.8, the correlation coefficient in CP SAR mode can be defined as [6]:

$$Corr = \frac{\text{Re}\{-i\langle E_H E_V^* \rangle\}}{\sqrt{\langle |E_H|^2 \rangle \langle |E_V|^2 \rangle}} \tag{29}$$

Additionally, for CP SAR modes, the coherency coefficient can be derived by:

$$Coh = \frac{|D_{12}|}{\sqrt{D_{11} D_{22}}} \tag{30}$$

where the coherency matrix $D$ for dual- and compact polarimetric SAR modes can be defined as:

$$D = \begin{pmatrix} \langle E_H + iE_V \rangle^2 & \langle E_H + iE_V \rangle \langle E_H - iE_V \rangle^* \\ \langle E_H + iE_V \rangle^* \langle E_H - iE_V \rangle & \langle E_H - iE_V \rangle^2 \end{pmatrix} \tag{31}$$

### 2.5. Supervised Classifications

Supervised classifications can take advantage of training data samples to set up the decision rule for classification, which has the best capability of fitting training datasets, as well as predicting the class of testing data samples. In this paper, three largely used supervised classifiers are considered.

### 2.5.1. Support Vector Machine (SVM)

SVM is based on structural risk minimization, the basic idea of which is to map multi-dimensional feature into a higher dimensional space and use a hyperplane to separate them linearly with the maximum margin between different classes [35]. SVM has superb performance in dealing with classification problems with a small number of training datasets. It firstly maps training vectors $x_i$ into a higher dimensional space by using kernel function $\Phi$ and, hence, finds a linear separating hyperplane with the maximal margin in this higher dimensional space. In this paper, the radial basis function is adopted as the kernel function.

### 2.5.2. Artificial Neural Network (ANN)

ANN was designed based on the nervous systems of animals [36]. It can be used to estimate the complicated unknown functions based on a large number of inputs. ANNs are often used for supervised classification for their adaptive nature. They can often obtain good performance when the training samples are sufficient. In this paper, the feed-forward neural network (FFNN) with three layers is considered. In the FFNN, each neuron (or call "unit") contains a transfer function. The neuron of the hidden and output units performs the nonlinear sigmoid function, while the input units have an identity transfer function. Then, layers are connected to each other by a system of weights, which multiplicatively scale the values traversing the links. The weights and bias of these links in the network is firstly randomly initiated and then fine-tuned through the backpropagation process.

### 2.5.3. Maximum Likelihood Classification (ML)

ML is a kind of classical classifier that is widely used in a variety of remote sensing applications. Based on training data, the maximum likelihood method selects the set of values of the model parameters that maximizes the likelihood function [37].

### 2.6. Features Selection Scheme

In a classification scheme, continuously adding features generates the well-known pattern recognition problem known as the "curse of dimensionality", which means that the classification performance will not always improve with the increase of added features, especially when the number of training data samples is limited. Sometimes, "bad" features may even largely lower the classification accuracy. Moreover, the increase of the number of features makes the classification algorithms time consuming. In this paper, a forward feature selection scheme is considered, to choose the optimum feature sets for each classifier: starting from the best 2 features, the classification chooses to add the feature that provides the largest improvement on classification accuracy at each time. Then, in the comprehensive analysis, feature sets that achieved the best classification performance are employed.

### 2.7. Classification Accuracy Evaluation

In this study, overall accuracy (OA) and kappa coefficients (*Kappa*) are employed to quantitatively evaluate the classification accuracy. They can be derived from the confusion matrix of the testing data samples, where the rows represent the classified results and columns represent the referenced data. In the confusion matrix, the last row is the sum of all previous rows, and the last column is the sum of all previous columns. The OA is calculated by summing the number of pixels classified correctly divided by the total number of pixels, and the kappa coefficient measures the accuracy of the classification in another way; the definitions of both of them are shown below:

$$OA = \frac{\sum_{i=1}^{n-1} X_{ii}}{X_{nn}} \tag{32}$$

$$Kappa = \frac{X_{nn} \sum_{i=1}^{n-1} X_{ii} - \sum_{i=1}^{n-1} (X_{in} X_{ni})}{X_{nn}^2 - \sum_{i=1}^{n-1} (X_{in} X_{ni})} \tag{33}$$

where $X = \{x_{ij}\}_{n \times n}$ $(i, j = 1, 2, 3, \dots, n)$ is the confusion matrix and $X_{in}$ stands for the number of samples that belongs to the *i*-th class and classified as *n*-th class.

### 2.8. Dimension Reduction

Various features can be extracted from polarimetric SAR data. However, inevitably, they are correlated and suffer from noise. In this study, three typical algorithms, principle component analysis (PCA), local linear embedding (LLE) and ISOMAP, were comparatively employed to reduce the dimension of polarimetric SAR features.

PCA reduces the number of features by replacing them with their linear combination. These new features are derived by the idea of maximizing their variance and making them uncorrelated. It comes from the theory of linear algebra; PCA has been abundantly used in many applications and has become a very popular method for its highly efficient, non-parametric characteristic.

LLE is an unsupervised learning algorithm that computes low-dimensional, neighborhood-preserving embeddings of high-dimensional inputs. It maps its inputs into a single global coordinate system of lower dimensionality, and its optimizations do not involve local minima. LLE is capable of learning the global structure of nonlinear manifolds based on the exploration of the local symmetries of linear reconstructions [38].

ISOMAP takes advantage of local metric information by measuring geodesic distances and learning the underlying global geometry of a dataset. Developed from multidimensional scaling (MDS), it is capable of discovering the nonlinear degrees of freedom that underlie complex natural observations, such as human handwriting or images of a face under different viewing conditions [39].

### 3. Results

In this study, features extracted from RADARSAT-2 quad-pol SAR data were analyzed. The pseudo RGB image of the Radarsat-2 data on the Pauli basis are provided in Figure 2. It was acquired during the 2011 Norwegian oil-on-water experiment, in which three verified slicks were present; from left to right, they were: biogenic film, emulsions and mineral oil [34]. The biogenic film was simulated by Radiagreen ebo plant oil. Emulsions were made of Oseberg blend crude oil mixed with 5% IFO380 (Intermediate Fuel Oil), released 5 h before the radar acquisition. Additionally, the Balder crude oil was released 9 h before the radar acquisition [34].

In this study, the effect of feature numbers on the final classification result is analyzed, by considering three major supervised classifiers, namely, SVM, ANN and ML. Based on the quad-pol SAR data, dual-pol and compact polarimetric SAR data were also simulated, then features were extracted based on uniform feature extraction algorithms. Before the process of supervised classification, all of the features were normalized to the range of 0–1. Finally, the performance of features extracted from different polarimetric SAR modes in oil spill classification are compared and analyzed.

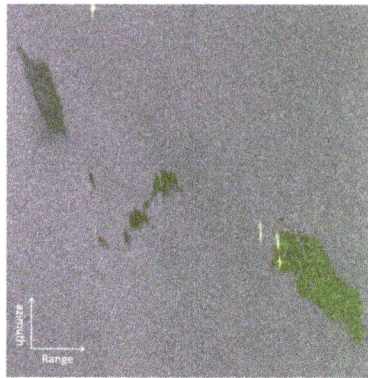

**Figure 2.** Pauli RGB image of RADARSAT-2 data.

In the supervised classification experiment, 5393 and 5467 pixels of mineral oil covered and non-covered (including clean sea surface and biogenic films) training samples were picked within the study area respectively. Then, 5550 and 5535 testing samples of these two types are picked as the ground truth. The training and testing samples do not include each other. Both the training and testing sample include comparable numbers of pixels that are visually identified (based on ground truth) as clean sea surface, mineral oil and biogenic films (weak-damping surfactants).

*3.1. Oil Spill Classification Based on Fully Polarimetric SAR Features*

In the classification based on quad-pol SAR data, feature numbers from 2–10 are considered. The polarimetric features derived from quad-pol SAR data considered in the study are listed in Table 3. All of the features considered in this experiment are provided in Figure 3. In the display, all of the features are normalized to [0, 1]. In Figure 4, the tendency of overall accuracy achieved by three classifiers is plotted. The best classification result was achieved when considering eight features for SVM, nine features for ANN and four features for MLC. Generally, SVM achieved the best classification performance, followed by ANN. This result proved the superb capability of SVM in dealing with a large number of features. It can be observed that in all of the classifications, after the best four features have been introduced, the classification results began to fluctuate and did not change very much. These four features are: pedestal height, correlation coefficient, standard deviation of CPD and alpha angle. The eight features used for SVM classification are: $S_{VV}^2$, pedestal height, entropy, $DoP_{HHVV}$, correlation coefficient, coherency coefficient, standard deviation of CPD and alpha angle. The nine features used for ANN are all of the features except ellipticity. As introduced in the previous session, all of these features have strong physical meaning, which enables them to largely contribute to the classification between mineral oil and clean sea surface/biogenic film. They are also not likely affected by the noise floor.

**Table 3.** Features that derived from quad-polarimetric SAR data.

| Number | Feature |
|--------|---------|
| 1 | $S_{VV}^2$ |
| 2 | Pedestal Height |
| 3 | Entropy |
| 4 | $DoP_{HHVV}$ |
| 5 | Correlation Coefficient |
| 6 | Conformity Coefficient |
| 7 | Coherency Coefficient |
| 8 | Ellipticity $\chi$ |
| 9 | CPD Standard Deviation |
| 10 | Alpha Angle |

Note: Features 4 and 8 were extracted from the Stokes vector considering the *HH* and *VV* channels.

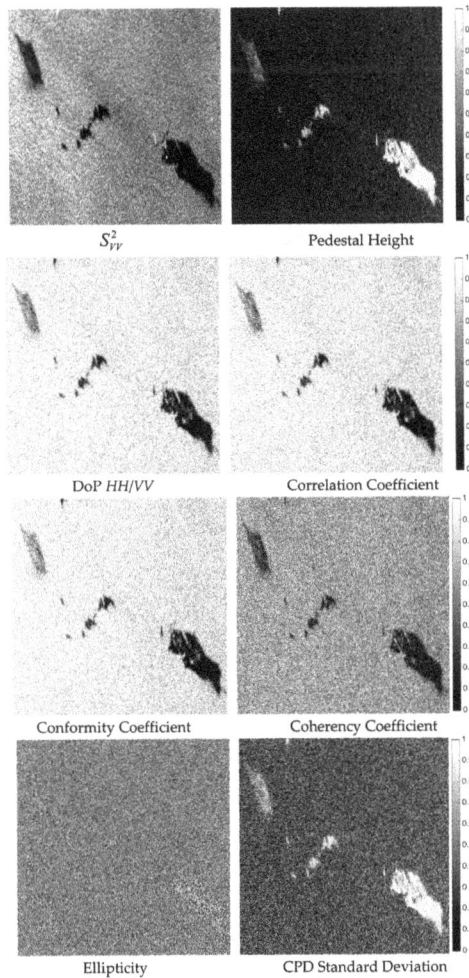

**Figure 3.** *Cont.*

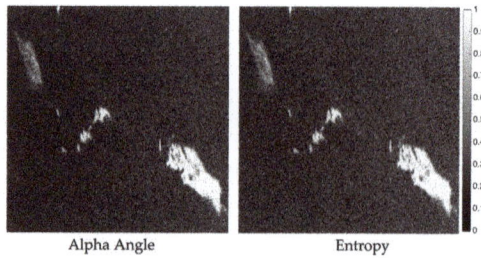

Alpha Angle                         Entropy

**Figure 3.** Quad-pol features extracted from the RASARSAT-2 data.

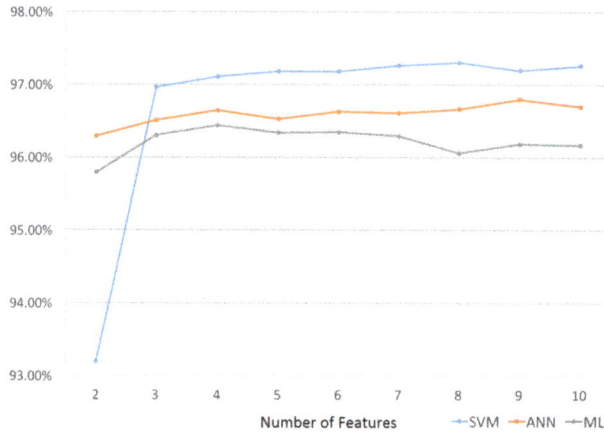

**Figure 4.** Classification accuracy achieved by three classifiers with the number of features changing from 2–10.

The best classification result was achieved by SVM with eight quad-polarimetric SAR features. This is shown in Figure 5a. Figure 5b,c demonstrates the classification results obtained by ML and ANN, respectively, where the red color indicates mineral oil and green indicates non-oil area. The confusion matrix of the best classification results achieved by these three classifiers is listed in Tables 4–6. From the detailed analysis on the confusion matrix of these classification results, it can be observed that the major reason that SVM is superior to the other two classifiers is that it successfully controlled the commission error of non-oil area, namely the error caused by wrongly classified clean sea surface and biogenic slicks.

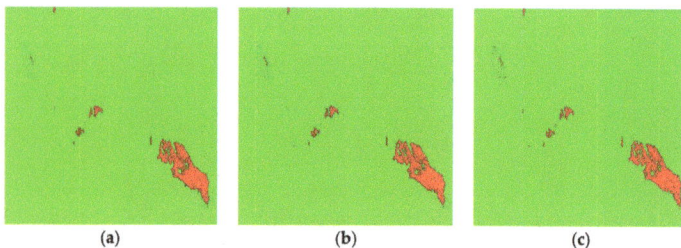

(a)                         (b)                         (c)

**Figure 5.** Classification results based on quad-pol SAR features using different classifiers. (**a**) SVM; (**b**) ML; (**c**) ANN.

**Table 4.** Confusion matrix achieved by SVM based on 8 fully polarimetric features.

| Class | Ground Truth (Pixels) | | |
|---|---|---|---|
| | Oil | Sea | Total |
| Oil | 5429 | 178 | 5607 |
| Sea | 121 | 5357 | 5478 |
| Total | 5550 | 5535 | 11,085 |

Overall accuracy = 97.3027% (10,786/11,085), kappa coefficient = 0.9461.

**Table 5.** Confusion matrix achieved by ML based on 4 fully polarimetric features.

| Class | Ground Truth (Pixels) | | |
|---|---|---|---|
| | Oil | Sea | Total |
| Oil | 5411 | 256 | 5667 |
| Sea | 139 | 5279 | 5418 |
| Total | 5550 | 5535 | 11,085 |

Overall accuracy = 96.4366% (10,690/11,085), kappa coefficient = 0.9287.

**Table 6.** Confusion matrix achieved by ANN based on 9 fully polarimetric features.

| Class | Ground Truth (Pixels) | | |
|---|---|---|---|
| | Oil | Sea | Total |
| Oil | 5427 | 232 | 5659 |
| Sea | 123 | 5303 | 5426 |
| Total | 5550 | 5535 | 11,085 |

Overall accuracy = 96.7975% (10,730/11,085), kappa coefficient = 0.9359.

### 3.2. Oil Spill Classification Based on Different Polarimetric SAR Modes

In this part, as listed in Table 7, dual- and compact polarimetric SAR features are extracted from simulated SAR datasets (the conformity coefficient is only available in $\pi/2$ mode). The overall classification accuracy of three classifiers based on the features extracted from different polarimetric SAR modes is compared in Figure 6.

**Table 7.** Uniform dual and compact polarimetric features considered in the study.

| Number | Feature * |
|---|---|
| 1 | $E_V^2$ |
| 2 | Pedestal Height (CP) |
| 3 | Entropy (CP) |
| 4 | DoP (CP) |
| 5 | Correlation Coefficient (CP) |
| 6 | Alpha Angle (CP) |
| 7 | Coherency Coefficient (CP) |
| 8 | Ellipticity $\chi$ (CP) |
| 9 | CPD Standard Deviation (CP) |
| 10 | Conformity Coefficient ($\pi/2$) |

* Features 1–9 are extracted from dual and compact polarimetric SAR data following the methods introduced in Section 2.4, while Feature 10 is only available for $\pi/2$ mode. "CP" stands for features derived from compact polarimetric SAR data in order to distinguish them from those calculated from quad-pol SAR data.

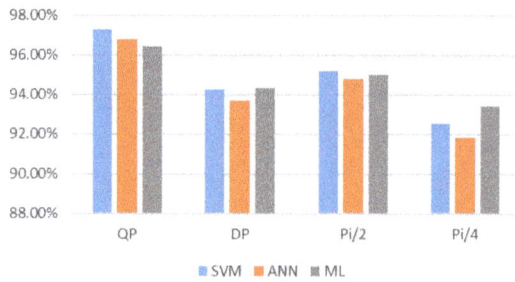

**Figure 6.** Classification accuracy of different polarimetric SAR modes achieved by SVM, ANN and ML.

Quad-pol (QP) feature-based classification has the highest OA, followed by $\pi/2$ compact polarimetric SAR mode and $HH/VV$ dual-polarized (DP) mode. $\pi/4$ mode-based classification has the lowest performance. In QP and $\pi/2$ modes, SVM achieved the best performance, while for $HH/VV$ DP and $\pi/4$ modes, better performance was achieved by ML. Furthermore in dual- and compact polarimetric SAR modes, ML outperformed ANN; this may be explained by the fact that ANN has a higher requirement to the separability of the dataset and is more vulnerable to the loss or mixture of crucial information of the dataset. The confusion matrices of the classification results achieved by SVM based on features extracted from different polarimetric SAR modes are listed in Tables 8–10, with the feature number that achieved the best classification performance, and the classification results are demonstrated in Figure 7a–c.

Similar supervised classification experiments were also conducted based on single polarimetric feature $S_{VV}^2$ only. A much lower overall accuracy (61.7772%) and kappa coefficient (0.2348) were obtained. Figure 7d shows the classification result, from which it could be observed that most parts of the biogenic slick were misclassified to mineral oil. The confusion matrix (Table 11) further supported this observation. This result manifested the limitation of single polarimetric SAR mode in distinguishing mineral oil and biogenic films.

**Table 8.** Confusion matrix achieved by SVM based on 9 dual-polarized (DP) mode features.

| Class | Ground Truth (Pixels) | | |
|---|---|---|---|
| | Oil | Sea | Total |
| Oil | 5357 | 445 | 5802 |
| Sea | 193 | 5090 | 5283 |
| Total | 5550 | 5535 | 11,085 |

Overall accuracy = 94.2445% (10,447/11,085), kappa coefficient = 0.8849.

**Table 9.** Confusion matrix achieved by SVM based on $10\pi/2$ mode features.

| Class | Ground Truth (Pixels) | | |
|---|---|---|---|
| | Oil | Sea | Total |
| Oil | 5378 | 363 | 5741 |
| Sea | 172 | 5172 | 5344 |
| Total | 5550 | 5535 | 11,085 |

Overall accuracy = 95.1737% (10,550/11,085), kappa coefficient = 0.9035.

**Table 10.** Confusion matrix achieved by SVM based on $9\pi/4$ mode features.

| Class | Ground Truth (Pixels) | | |
|---|---|---|---|
| | Oil | Sea | Total |
| Oil | 5316 | 595 | 5911 |
| Sea | 234 | 4940 | 5174 |
| Total | 5550 | 5535 | 11,085 |

Overall accuracy = 92.5214% (10,256/11,085), kappa coefficient = 0.8504.

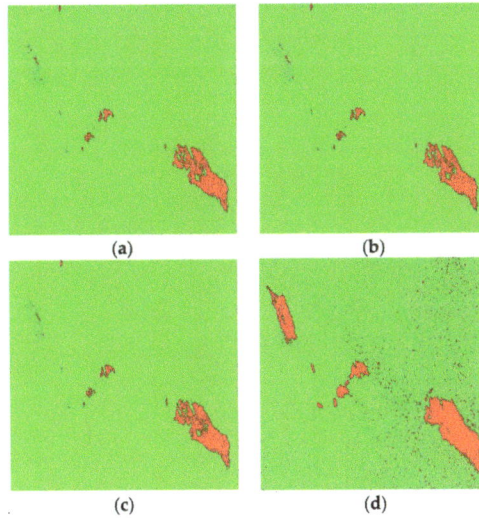

**Figure 7.** Classification result using SVM based on the features of: (**a**) DP mode; (**b**) $\pi/4$ mode; (**c**) $\pi/2$ mode; (**d**) $S_{VV}^2$.

**Table 11.** Confusion matrix achieved by SVM based on $S_{VV}^2$.

| Class | Ground Truth (Pixels) | | |
|---|---|---|---|
| | Oil | Sea | Total |
| Oil | 5438 | 4125 | 9563 |
| Sea | 112 | 1410 | 1522 |
| Total | 5550 | 5535 | 11,085 |

Overall accuracy = 61.7772% (6848/11,085), kappa coefficient = 0.2348.

### 3.3. Oil Spill Classification Based on Dimension Reduction of Features

Based on the new feature sets, classification was conducted by using SVM. The classification results obtained by employing the three feature dimension reduction methods are shown in Figure 8. Tables 12–14 demonstrate the performance of classification. The feature set derived from LLE achieved the highest overall accuracy of 92.1696%. The feature set derived from PCA obtained an OA of 91.1322%, with the lowest false alarm rate. The feature set derived from ISOMAP had an OA of 90.8705%, which is the lowest among these three algorithms. Generally, feature reduction algorithms have acceptable performance in keeping the key information for distinguishing mineral oil and biogenic films. However, in this experiment, the performance achieved by dimension reduced feature sets is constantly lower than the original feature sets, which may be related to the issue of sample selection.

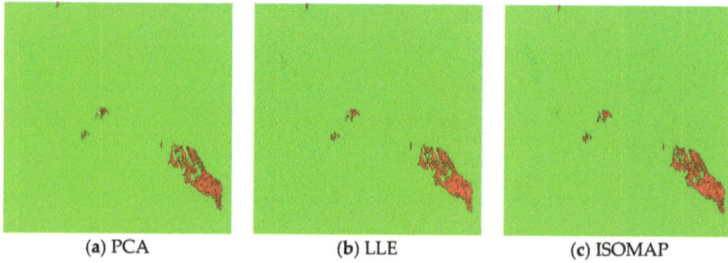

(a) PCA        (b) LLE        (c) ISOMAP

**Figure 8.** Classification results using SVM based on feature dimension reduction methods.

**Table 12.** Confusion matrix achieved by SVM based on four features derived from PCA on quad-pol SAR features.

| Class | Ground Truth (Pixels) | | |
|-------|------|------|-------|
| | Oil | Sea | Total |
| Oil | 4649 | 82 | 4731 |
| Sea | 901 | 5453 | 6354 |
| Total | 5550 | 5535 | 11,085 |

Overall accuracy = 91.1322% (10,102/11,085), kappa coefficient = 0.8227.

**Table 13.** Confusion matrix achieved by SVM based on four features derived from local linear embedding (LLE) on quad-pol SAR features.

| Class | Ground Truth (Pixels) | | |
|-------|------|------|-------|
| | Oil | Sea | Total |
| Oil | 4879 | 197 | 5076 |
| Sea | 671 | 5338 | 6009 |
| Total | 5550 | 5535 | 11,085 |

Overall accuracy = 92.1696% (10,217/11,085), kappa coefficient = 0.8434.

**Table 14.** Confusion matrix achieved by SVM based on four features derived from ISOMAP on quad-pol SAR features.

| Class | Ground Truth (Pixels) | | |
|-------|------|------|-------|
| | Oil | Sea | Total |
| Oil | 4809 | 271 | 5080 |
| Sea | 741 | 5264 | 6005 |
| Total | 5550 | 5535 | 11,085 |

Overall accuracy = 90.8705% (10,073/11,085), kappa coefficient = 0.8174.

## 4. Discussion

With the help of polarimetric information, oil slicks and their biogenic films can be well separated. Experiments proved that the classification performance does not always increase with introducing more features; it fluctuates or decreases after the sufficient features are considered. This effect can be attributed to correlated and contradicting information carried in these features. In the demonstrated case, a set of four key features is sufficient, and the classification performance does not increase much when introducing more features. This phenomenon shows that most polarimetric information can be provided by several powerful and complementary features. As a result, in real applications, only a few representative features need to be extracted to save computing time and avoid the problem of "curse of dimensionality".

In this study, we present a comparative study on features extracted from different polarimetric SAR modes to provide valuable information for oil spill classification. It was proven that quad-pol features have the highest overall accuracy, while $\pi/2$ compact polarimetric SAR modes had the best performance among all compact and dual-polarimetric SAR modes, followed by $HH/VV$ dual-polarimetric SAR modes. The lowest performance was achieved by $\pi/4$ mode. In $\pi/2$ mode, the circularly-polarized signal is transmitted, which has been proven to be more suitable for a series of marine remote sensing applications [6,23], since it is very sensitive to the change of scattering mechanisms on the sea surface. *HH-VV* phase correlation is very helpful for distinguishing marine oil spill and biogenic oil slicks [3], and thus, *HH/VV* dual-polarization mode achieved relatively good performance.

In fully and $\pi/2$ compact polarimetric modes when the separability of the features is high, SVM achieved the highest performance in comparison with other supervised classifiers. The advantage of SVM is its good capability of handling the problem of the "curse of dimensionality". It has better performance in dealing with data of a high dimensional feature space in supervised classification applications, such as this illustrated case. For quad-pol feature-based classification, ANN performed slightly better than ML, and for other modes, ML performed better than ANN. A possible explanation is that ANN is very sensitive to the quality of features and has the trend of over-training when dealing with features with disturbance. Therefore, in compact and dual-polarimetric SAR modes, ML performs better than ANN, although the latter one is more sophisticated in its architecture.

This study shows that polarimetric SAR can distinguish mineral oil from biogenic slicks. An important result is that the identification of different oils (bunker oil, crude oil, petrochemical films) is very important for clean-up operations. Different oils have different physical/chemical properties, e.g., viscosity, density, evaporation rate, etc., and theoretically, a difference in these properties can be observed in polarimetric SAR images. However, currently, there is not enough valid data to support this latter postulate. This analysis can be made in the future.

It is important to analyze the behavior of weathering oil in polarimetric SAR images. Particularly, evaporation, emulsification and sinking are important related slick detections by SAR. Studies [40,41] indicate that the percentages of oil trapped, evaporated and at the surface vary with the type of oil spilt and with the location in which spills are firstly generated. In essence, the movement of oil, its original type/density and the time that leads to its emulsification/evaporation/sinking are variable in different oil spills. It is also considered crucial to understand the effects of emulsification and ocean-driven slick movement in the size(s) and distribution of oil slicks at the surface for environmental protection [42]. Hence, more detailed experiments should be made to quantitatively analyze the degree of degradation of an oil spill based on polarimetric SAR.

## 5. Conclusions

The Norwegian oil-on-water experiment in 2011 provided polarimetric SAR acquisition with verified oil spill and biogenic slicks on one scene of Radarsat-2 data. More quad-pol SAR data samples are being further collected to derive more detailed and convincing results in the near future studies.

The key findings of this comparative study can be summarized as follows:

- Polarimetric SAR features can be input into supervised algorithms to achieve reliable oil spill classification. For this dataset, a feature set with four features is sufficient for most polarimetric features based oil spill classifications. They are: pedestal height, correlation coefficient, standard deviation of CPD and alpha angle.
- Among all of the compact polarimetric SAR modes, $\pi/2$ mode has the best performance among all of the dual- and compact polarimetric SAR modes, for its sensitivity to different scattering mechanisms caused by mineral oil and biogenic look-alikes.

- Among all of the supervised classifiers, SVM outperforms other classifiers when sufficient polarimetric information can be obtained, such as quad-pol mode. ML performs better than other supervised classifiers when only incomplete polarimetric information is available, such as traditional dual-pol and $\pi/4$ mode.

The reasons for the unreliable results in feature reduction experiments may be attributed to insufficient data sampling when computing feature maps. The understanding of oil in the characteristics of polarimetric SAR imagery is key to optimize the processing procedures of automatic oil spill detection and classification algorithms.

In the near future, there will be more compact polarimetric SAR data available for marine surveillance applications. The polarimetric observation capabilities of these sensors will largely improve the efficiency and reliability of oil spill detection and any future classifications applications based on SAR data.

**Acknowledgments:** The SAR data of Radarsat-2 is highly appreciated. This research is jointly supported the "2015 Innovation Programs for Research and Entrepreneurship Teams in Jiangsu Province, China", the National Key Research and Development Program of China (2016YFC1402003) and the Priority Academic Program Development of Jiangsu Higher Education Institutions (PAPD).

**Author Contributions:** Yuanzhi Zhang and Yu Li conceived of and designed the experiments. Yu Li performed the experiments. Yuanzhi Zhang, Yu Li, X. San Liang and Jinyeu Tsou analyzed the data and wrote the paper.

**Conflicts of Interest:** The authors declare no conflict of interest. The founding sponsors had no role in the design of the study; in the collection, analyses or interpretation of data; in the writing of the manuscript; nor in the decision to publish the results.

## References

1. Alpers, W.; Espedal, H. Oils and surfactants. In *Synthetic Aperture Radar Marine User's Manual*; US Department of Commerce: Washington, DC, USA, 2004; pp. 263–275.
2. Migliaccio, M.; Gambardella, A.; Tranfaglia, M. SAR Polarimetry to Observe Oil Spills. *IEEE Trans. Geosci. Remote Sens.* **2007**, *45*, 506–511. [CrossRef]
3. Migliaccio, M.; Nunziata, F.; Gambardella, A. On the co-polarized phase difference for oil spill observation. *Int. J. Remote Sens.* **2009**, *30*, 1587–1602. [CrossRef]
4. Topouzelis, K.; Stathakis, D.; Karathanassi, V. Investigation of Genetic Algorithms Contribution to Feature Selection for Oil Spill Detection. *Int. J. Remote Sens.* **2008**, *30*, 611–625. [CrossRef]
5. Marghany, M.; Hashim, M. Discrimination between oil spill and look-alike using fractal dimension algorithm from RADARSAT-1 SAR and AIRSAR/POLSAR data. *Int. J. Phys. Sci.* **2011**, *6*, 1711–1719.
6. Zhang, B.; Perrie, W.; Li, X.; Pichel, W. Mapping sea surface oil slicks using RADARSAT-2 quad-polarization SAR image. *Geophys. Res. Lett.* **2011**, *38*, 415–421. [CrossRef]
7. Zhang, Y.; Lin, H.; Liu, Q.; Hu, J.; Li, X.; Yeung, K. Oil-spill monitoring in the coastal waters of Hong Kong and vicinity. *Mar. Geod.* **2012**, *35*, 93–106. [CrossRef]
8. Guo, J.; He, Y.; Long, X.; Hou, C.; Liu, X.; Meng, J. Repair wind field in oil contaminated areas with SAR images. *Chin. J. Oceanol. Limnol.* **2015**, *33*, 525–533. [CrossRef]
9. Gade, M.; Alpers, W. Using ERS-2 SAR for routine observation of marine pollution in European coastal waters. *Sci. Total Environ.* **1999**, *237*, 38441–38448. [CrossRef]
10. Suri, S.; Schwind, P.; Uhl, J.; Reinartz, P. Modification in the SIFT operator for effective SAR image matching. *Int. J. Image Data Fusion* **2010**, *1*, 243–256. [CrossRef]
11. Minchew, B.; Jones, C.E.; Holt, B. Polarimetric Analysis of Backscatter from the Deepwater Horizon Oil Spill Using L-Band Synthetic Aperture Radar. *IEEE Trans. Geosci. Remote Sens.* **2012**, *50*, 3812–3830. [CrossRef]
12. Migliaccio, M.; Nunziata, F.; Buono, A. SAR polarimetry for sea oil slick observation. *Int. J. Remote Sens.* **2015**, *36*, 3243–4273. [CrossRef]
13. Li, H.; Perrie, W.; He, Y.; Wu, J.; Luo, X. Analysis of the Polarimetric SAR Scattering Properties of Oil-Covered Waters. *IEEE J. Sel. Top. Appl. Earth Obs. Remote Sens.* **2015**, *8*, 3751–3759. [CrossRef]
14. Nunziata, F.; Migliaccio, M. Gambardella, A. Pedestal height for sea oil slick observation. *IET Radar Sonar Navig.* **2010**, *5*, 103–110. [CrossRef]

15. Alpers, W. Remote sensing of oil spills. In Proceedings of the Symposium Maritime Disaster Management, King Fahd University of Petroleum and Minerals, Dhahran, Saudi Arabia, 19–23 January 2002; pp. 19–23.

16. Souyris, J.C.; Imbo, P.; Fjortoft, R.; Mingot, S.; Lee, J.-S. Compact polarimetry based on symmetry properties of geophysical media: The $\pi/4$ mode. *IEEE Trans. Geosci. Remote Sens.* **2005**, *43*, 634–646. [CrossRef]

17. Sabry, R.; Vachon, P.W. A Unified Framework for General Compact and Quad Polarimetric SAR Data and Imagery Analysis. *IEEE Trans. Geosci. Remote Sens.* **2014**, *52*, 582–602. [CrossRef]

18. Chen, J.; Quegan, S. Calibration of Spaceborne CTLR Compact Polarimetric Low-Frequency SAR Using Mixed Radar Calibrators. *IEEE Trans. Geosci. Remote Sens.* **2011**, *49*, 2712–2723. [CrossRef]

19. Nord, M.E.; Ainsworth, T.L.; Lee, J.-S.; Stacy, N.J.S. Comparison of Compact Polarimetric Synthetic Aperture Radar Modes. *IEEE Trans. Geosci. Remote Sens.* **2009**, *47*, 174–188. [CrossRef]

20. Dubois-Fernandez, P.C.; Souyris, J.-C.; Angelliaume, S.; Garestier, F. The Compact Polarimetry Alternative for Spaceborne SAR at Low Frequency. *IEEE Trans. Geosci. Remote Sens.* **2008**, *46*, 3208–3222. [CrossRef]

21. Collins, M.J.; Denbina, M.; Atteia, G. On the Reconstruction of Quad-Pol SAR Data from Compact Polarimetry Data for Ocean Target Detection. *IEEE Trans. Geosci. Remote Sens.* **2013**, *51*, 591–600. [CrossRef]

22. Li, H.; Perrie, W.; He, Y.; Lehner, S.; Brusch, S. Target Detection on the Ocean with the Relative Phase of Compact Polarimetry SAR. *IEEE Trans. Geosci. Remote Sens.* **2013**, *5*, 3299–3305. [CrossRef]

23. Yin, J.; Yang, J.; Zhang, X. On the ship detection performance with compact polarimetry. In Proceedings of the 2011 IEEE Radar Conference (RADAR), Kansas City, MO, USA, 23–27 May 2011; pp. 675–680.

24. Raney, R.K. Hybrid-Polarity SAR Architecture. *IEEE Trans. Geosci. Remote Sens.* **2007**, *45*, 3397–3404. [CrossRef]

25. Shirvany, R.; Chabert, M.; Tourneret, J.-Y. Ship and Oil-Spill Detection Using the Degree of Polarization in Linear and Hybrid/Compact Dual-Pol SAR. *IEEE J. Sel. Top. Appl. Earth Obs. Remote Sens.* **2012**, *5*, 885–892. [CrossRef]

26. Cloude, S.R.; Goodenough, D.G.; Chen, H. Compact Decomposition Theory. *IEEE Geosci. Remote Sens. Lett.* **2012**, *9*, 28–32. [CrossRef]

27. Salberg, A.-B.; Rudjord, O.; Solberg, A.H.S. Oil Spill Detection in Hybrid-Polarimetric SAR Images. *IEEE Trans. Geosci. Remote Sens.* **2014**, *52*, 6521–6533. [CrossRef]

28. Nunziata, F.; Migliaccio, M.; Li, X. Sea Oil Slick Observation Using Hybrid-Polarity SAR Architecture. *IEEE J. Ocean. Eng.* **2014**, *1*, 426–440. [CrossRef]

29. Yin, J.; Yang, J.; Zhou, Z.; Song, J. The Extended Bragg Scattering Model-Based Method for Ship and Oil-Spill Observation Using Compact Polarimetric SAR. *IEEE J. Sel. Top. Appl. Earth Obs. Remote Sens.* **2014**, *99*, 1–13. [CrossRef]

30. Ulaby, F.T.; Moore, R.K.; Fung, A.K. *Microwave Remote Sensing. Active and Passive*; Artech House Inc.: Norwood, MA, USA, 1986; Volume 3.

31. Cloude, S.R.; Pottier, E. An entropy based classification scheme for land applications of polarimetric SAR. *IEEE Trans. Geosci. Remote Sens.* **1997**, *35*, 68–78. [CrossRef]

32. Shirvany, R. Estimation of the Degree of Polarization in Polarimetric SAR Imagery: Principles & Applicaions. Ph.D. Thesis, Institut National Polytechnique de Toulouse, Toulouse, France, October 2012.

33. Migliaccio, M.; Nunziata, F. On the exploitation of polarimetric SAR data to map damping properties of the Deepwater Horizon oil spill. *Int. J. Remote Sens.* **2014**, *35*, 3499–3519. [CrossRef]

34. Skrunes, S.; Brekke, C.; Eltoft, T. Characterization of Marine Surface Slicks by Radarsat-2 Multipolarization Features. *IEEE Trans. Geosci. Remote Sens.* **2014**, *52*, 5302–5319. [CrossRef]

35. Vapnik, V. *Statistical Learning Theory*; Wiley: New York, NY, USA, 1998.

36. Kavzoglu, T.; Mather, P.M. The use of backpropagating artificial neural networks in land cover classification. *Int. J. Remote Sens.* **2003**, *24*, 4907–4938. [CrossRef]

37. Ahmad, A.; Quegan, S. Analysis of Maximum Likelihood Classification on Multispectral Data. *Appl. Math. Sci.* **2012**, *6*, 6425–6436.

38. Roweis, S.T.; Saul, L.K. Nonlinear dimensionality reduction by locally linear embedding. *Science* **2000**, *290*, 2323–2326. [CrossRef] [PubMed]

39. Tenenbaum, J.B.; de Silva, V.; Langford, J.C. A global geometric framework for nonlinear dimensionality reduction. *Science* **2000**, *290*, 2319–2322. [CrossRef] [PubMed]

40. Alves, T.M.; Kokinou, E.; Zodiatis, G.; Radhakrishnan, H.; Panagiotakis, C.; Lardner, R. Multidisciplinary oil spill modelling to protect coastal communities and the environment of the Eastern Mediterranean Sea. *Sci. Rep.* **2016**, *6*, 36882. [CrossRef] [PubMed]
41. Alves, T.M.; Kokinou, E.; Zodiatis, G. A three-step model to assess shoreline and offshore susceptibility to oil spills: The South Aegean (Crete) as an analogue for confined marine basins. *Mar. Pollut. Bull.* **2014**, *86*, 443–457. [CrossRef] [PubMed]
42. Alves, T.M.; Kokinou, E.; Zodiatis, G.; Lardner, R. Hindcast, GIS and susceptibility modelling to assist oil spill clean-up and mitigation on the southern coast of Cyprus (Eastern Mediterranean). *Deep Sea Res. II Top. Stud. Oceanogr.* **2015**, *133*, 159–175. [CrossRef]

*applied*
*sciences*

MDPI

Article

# A TSVD-Based Method for Forest Height Inversion from Single-Baseline PolInSAR Data

Dongfang Lin, Jianjun Zhu *, Haiqiang Fu, Qinghua Xie and Bing Zhang

School of Geosciences and Info-Physics, Central South University, Changsha 410083, China;
lindongfang223@163.com (D.L.); haiqiangfu@csu.edu.cn (H.F.); csuxqh@126.com (Q.X.);
zhb210921@csu.edu.cn (B.Z.)
* Correspondence: zjj@csu.edu.cn; Tel.: +86-731-8883-6931

Academic Editors: Carlos López-Martínez and Juan Manuel Lopez-Sanchez
Received: 28 January 2017; Accepted: 21 April 2017; Published: 25 April 2017

**Abstract:** The random volume over ground (RVoG) model associates vegetation vertical structure parameters with multiple complex interferometric coherence observables. In this paper, on the basis of the RVoG model, a truncated singular value decomposition (TSVD)-based method is proposed for forest height inversion from single-baseline polarimetric interferometric synthetic aperture radar (PolInSAR) data. In addition, in order to improve the applicability of TSVD for this issue, a new truncation method is proposed for TSVD. Differing from the traditional three-stage method, the TSVD-based inversion method estimates the pure volume coherence directly from the complex interferometric coherence, and estimates the forest height from the estimated pure volume coherence with a least-squares method. As a result, the TSVD-based method can adjust the contributions of the polarizations in the estimation of the model parameters and avoid the null ground-to-volume ratio assumption. The simulated experiments undertaken in this study confirmed that the TSVD-based method performs better than the three-stage method in forest height inversion. The TSVD-based method was also applied to E-SAR P-band data acquired over the Krycklan Catchment, Sweden, which is covered with mixed pine forest. The results showed that the TSVD-based method improves the root-mean-square error by 48.6% when compared to the three-stage method, which further validates the performance of the TSVD-based method.

**Keywords:** polarimetric interferometric synthetic aperture radar (PolInSAR); vegetation height; truncated singular value decomposition (TSVD); least squares

## 1. Introduction

It is well known that vegetation height plays an important role in quantifying the terrestrial carbon cycle [1,2]. Moreover, vegetation height is an essential factor for the estimation of the biomass stored in vegetation [3,4]. Therefore, accurately extracting vegetation height at a large scale is an important task. Given the fact that polarimetric interferometric synthetic aperture radar (PolInSAR) can separate the scattering power of a single resolution cell into the contributions of surface, double-bounce, and volume scattering, it can be considered to be a viable remote sensing technique for estimating vegetation height in large-scale areas [4–9].

PolInSAR can be used to extract vegetation height through its sensitivity to the vegetation vertical structure [10–14]. The complex interferometric coherence of the observed PolInSAR data has been related to the vertical distribution of the vegetation scattering [10–12]. In a number of PolInSAR campaigns, the random volume over ground (RVoG) model [6,10] has been used to extract vegetation height from the complex interferometric coherence [15–17]. The RVoG model is a physical PolInSAR model that integrates the complex coherence and the biophysical parameters. Based on the RVoG model, Papathanassiou [6] proposed six-dimensional nonlinear optimization method, which has been successfully evaluated with different types of PolInSAR data [6,15,16]. However, for this method, the accuracy of the solution is greatly dependent on the selected initial value, and a poor choice of initial value can result in unstable parameter estimation. Furthermore, the iterative procedure used in this method consumes too much time and is unsuitable for the inversion of large-scale areas. To cope with this problem, Cloude [18] separated the six-dimensional nonlinear parameter optimization process into three stages. This solution is known as the "three-stage method", and has been widely used in vegetation height extraction for its simple, universal, and time-saving properties [15,19]. However, the three-stage method assumes that there is at least one polarization channel without a ground scattering contribution (the null ground-to-volume ratio (GVR) assumption), and thus it is difficult to determine the polarization because the volume and ground scattering contributions are always mixed in all of the polarization channels due to the diverse penetration depths [20]. As a result, the pure volume coherence estimated by the three-stage method is often inaccurate. Furthermore, with the three-stage method, it is not possible to adjust the contributions of the interferometric coherence observations in the estimation of the RVoG model parameters.

The aim of this work is to address the limitations described above and extract accurate vegetation heights from single-baseline PolInSAR data. Based on the RVoG model, linear observation equations are first developed from the complex interferometric coherence observations by Taylor expansion, in order to combine all the available interferometric coherence observations. Next, since the coefficient matrix derived from the linear equations is ill-conditioned, in order to overcome this ill-posed problem, the proposed truncated singular value decomposition (TSVD)-based method is used to estimate the pure volume coherence. Furthermore, due to the fact that the ordinary truncation method for TSVD is not suitable for this issue, a more adaptive truncation method is proposed so as to improve the accuracy of the estimated pure volume coherence. Finally, the forest height is extracted from the estimated pure volume coherence by the use of a least-squares method.

This paper is structured as follows. The principle of the RVoG model is introduced and discussed in Section 2. Section 3 presents the TSVD-based method for the estimation of vegetation height from complex interferometric coherence observations. The validation of the experiments are presented in Section 4. The discussions are presented in Section 5. Finally, the conclusions are drawn in Section 6.

## 2. RVoG Model

The RVoG model is a physical model that associates vegetation vertical structure parameters with multiple complex interferometric coherence observables [10]. It is a basic and popular model for describing vegetation scenarios. The model depicts the vegetation layer as a volume with randomly oriented particles over an impenetrable ground surface. Without considering the temporal decorrelation, the complex interferometric coherence $\gamma(\omega)$ is expressed as [6]:

$$\gamma(\omega) = e^{i\varphi_0} \frac{\gamma_v + \mu(\omega)}{1 + \mu(\omega),} \tag{1}$$

where $\omega$ is the unitary polarization vector that defines the choice of scattering mechanism, $\varphi_0$ denotes the ground surface phase, $\mu(\omega)$ represents the ground-to-volume ratio (GVR) accounting for the polarization diversity, which varies with $\omega$, and $\gamma_v$ denotes the pure volume coherence, which is linked to the vegetation height and is expressed as [6]:

$$\gamma_v = \frac{2\sigma \left( e^{(2\sigma h_v / \cos\theta + i k_z h_v)} - 1 \right)}{(2\sigma + i k_z \cos\theta)(e^{(2\sigma h_v / \cos\theta)} - 1),} \tag{2}$$

where $\sigma$ denotes the mean extinction coefficient, $h_v$ denotes the vegetation height, $\theta$ is the incidence angle, and $k_z$ is the vertical wave number, which depends on the imaging geometry and wavelength [21]. $k_z$ is given as:

$$k_z = \frac{4\pi\Delta\theta}{\lambda\sin\theta,} \tag{3}$$

where $\lambda$ denotes the wavelength, and $\Delta\theta$ represents the incidence angle difference between the master and slave images.

According to Equation (1), the theoretical loci of the complex coherence sets of different polarizations follow a straight line in the complex plane, as shown in Figure 1 [18]. The figure also depicts the determination of the ground surface phase, which is applied in the three-stage method. However, the complex coherence observations no longer follows a common straight line in practice, due to the coherence fluctuations caused by all the possible decorrelations. In order to reconstruct the straight line accurately from noisy coherence sets, a line-fit approach based on least squares is proposed by the use of the following regression model [10,18]:

$$\text{Im}(\gamma(\omega)) = c\text{Re}(\gamma(\omega)) + d, \tag{4}$$

where Re() and Im() denote the real and imaginary operations, respectively, $c$ is the slope of the coherence line, and $d$ is the intersection point to the imaginary axis. Once more than two complex coherence observations are provided, the coherence line can be determined by the line-fit approach. The ground phase $\varphi_0$ can then be identified by the two intersections ($\varphi_1$ and $\varphi_2$) of the coherence line and the unit circle [18]. However, the least-squares criterion may be unstable if the ellipticity of the coherence sets is high or if the complex coherence observations are too discrete, due to the noise of the polarizations.

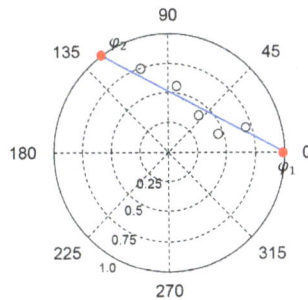

**Figure 1.** Geometrical interpretation of the coherence loci.

In addition, the three-stage method assumes that there is a polarization channel without ground scattering contribution, i.e., $\mu(\omega) = 0$, and then uses this assumption to further estimate the pure volume coherence. However, the volume and ground scattering contributions are always mixed in the polarization channels due to the diverse penetration depths. Therefore, it is difficult to find a polarization that can fit this hypothesis. As a result, when relying on this assumption, the estimated pure volume coherence will bias the extracted forest height.

As shown in Equation (1), $\gamma(\omega)$ is polarization-dependent. The complex coherence observations of the different scattering mechanisms can be obtained by the use of different unitary polarization vectors $\omega$. In other words, if the number of polarization vectors is $m$, then $m$ equations like Equation (1) can be established. By using the least-squares criterion, the unknown parameters in $\gamma(\omega)$ can be estimated when the number of parameters is less than that of the equations. Therefore, the least-squares method is a viable way to estimate vegetation height from complex coherence observations. The least-squares method directly estimates the RVoG model parameters from the complex coherence observations without any assumptions, and it is able to adjust the contributions of the coherence observations in the estimation of the model parameters.

### 3. The TSVD-Based Method for the Estimation of Vegetation Height From Complex Interferometric Coherence Observations

In this section, a novel approach is introduced for the estimation of vegetation height from single-baseline PolInSAR data, on the basis of the least-squares criterion and the TSVD-based method.

#### 3.1. Estimation of Pure Volume Coherence from the Complex Interferometric Coherence Observations

From Equations (1) and (2), it can be seen that $\gamma(\omega)$ and $\gamma_v$ are complex numbers. Therefore, the equations can be separated into two parts: a real part and an imaginary part. In order to simplify the nonlinear equations, we parameterize the pure volume coherence by $\gamma_v = a + bi$, and then the complex interferometric coherence can be given by:

$$\gamma(\omega) = e^{i\varphi_0}\frac{a + bi + \mu(\omega)}{1 + \mu(\omega)},\tag{5}$$

and the observational function for the least-squares criterion can be formulated as:

$$\gamma(\omega_j) = f(\varphi_0, a, b, \mu(\omega_j))j = 1, 2, \ldots, m,\tag{6}$$

where $\omega_j$ is the $j$-th unitary polarization vector, $f$ is the RVoG model function, as described in Equation (5), $\mu(\omega_j)$ is the $j$-th GVR associated with vector $\omega_j$, and $m$ represents the number of polarization projection vectors. It is clear that there are four unknown model parameters in each function, and $m + 3$ unknown parameters in $m$ functions.

To estimate the unknown model parameters, least-squares estimation is adopted to adjust the contributions of the complex coherence observations and suppress noise [20,22,23]. The corresponding least-squares criterion can be formulated as:

$$\sum_{j=1}^{m}|\hat{\gamma}(\omega_j) - \gamma(\omega_j)|^2 = \min,\tag{7}$$

where $\sum$ represents the summation operation, $|\cdot|$ represents the modulus operation, and $\hat{\gamma}(\omega_j)$ is the estimation of $\gamma(\omega_j)$. Since the complex coherence can be separated into a real part and an imaginary part, Equation (7) can be equally converted to:

$$\sum_{j=1}^{m}\left(|\text{Re}(\hat{\gamma}(\omega_j)) - \text{Re}(\gamma(\omega_j))|^2 + |\text{Im}(\hat{\gamma}(\omega_j)) - \text{Im}(\gamma(\omega_j))|^2\right) = \min.\tag{8}$$

In this way, the residual functions for the least-squares criterion are formulated as:

$$\begin{cases} V_{\text{Re}}^j = \text{Re}(\hat{\gamma}(\omega_j)) - \text{Re}(\gamma(\omega_j)), \\ V_{\text{Im}}^j = \text{Im}(\hat{\gamma}(\omega_j)) - \text{Im}(\gamma(\omega_j)), \end{cases}\tag{9}$$

where $V_{\text{Re}}^j$ denotes the residual of the real part corresponding to the $j$-th unitary polarization vector, and $V_{\text{Im}}^j$ denotes the residual of the imaginary part. From Equation (9), it can be seen that, if the number of polarization projection vectors is $m$, then $2m$ residual equations can be established. The number of unknown parameters is $m + 3$. Therefore, the unknowns can be estimated by least-squares estimation when the number of polarization projection vectors is more than three.

From Equation (5), it can be seen that the complex coherence function is highly nonlinear, which greatly affects the efficacy of the least-squares estimation. In view of this, a linearized strategy based on a Taylor series is adopted to convert the nonlinear function into a linear function [23]. In this case, the residual functions for the least-squares criterion can be rewritten as:

$$\begin{cases} V_{\text{Re}}^j = \dfrac{\partial \text{Re}(\gamma(\omega_j))}{\partial \varphi_0}\text{d}\varphi_0 + \dfrac{\partial \text{Re}(\gamma(\omega_j))}{\partial a}\text{d}a + \dfrac{\partial \text{Re}(\gamma(\omega_j))}{\partial b}\text{d}b + \dfrac{\partial \text{Re}(\gamma(\omega_j))}{\partial \mu(\omega_j)}\text{d}\mu(\omega_j) - l_{\text{Re}}^j \\ V_{\text{Im}}^j = \dfrac{\partial \text{Im}(\gamma(\omega_j))}{\partial \varphi_0}\text{d}\varphi_0 + \dfrac{\partial \text{Im}(\gamma(\omega_j))}{\partial a}\text{d}a + \dfrac{\partial \text{Im}(\gamma(\omega_j))}{\partial b}\text{d}b + \dfrac{\partial \text{Im}(\gamma(\omega_j))}{\partial \mu(\omega_j)}\text{d}\mu(\omega_j) - l_{\text{Im}}^j \end{cases} \tag{10}$$

where $\partial$ represents the partial derivative operation, $\text{d}\varphi_0$, $\text{d}a$, $\text{d}b$, and $\text{d}\mu(\omega_j)$ denote the corrections of the approximations of the unknown parameters, and $l_{\text{Re}}^j$ and $l_{\text{Im}}^j$ are the real and imaginary differences between the observed complex coherence and the predicted initial value. The vector $V$ denotes the residuals, $X$ denotes the unknown corrections, $A$ denotes the coefficient matrix, and $L$ denotes the real and imaginary differences. The residual functions for the least-squares criterion can then be expressed as [23,24]:

$$V = AX - L. \tag{11}$$

The unitary polarization vector can be constructed by linear-basis polarization, Pauli-basis polarization, magnitude diversity optimization polarization [6], and phase diversity polarization [25]. Therefore, more than three polarization projection vectors can be obtained for the estimation of the unknown parameters. Using the least-squares criterion, an estimate of the unknown corrections can be given by:

$$\hat{X} = \left(A^{\text{T}}A\right)^{-1}A^{\text{T}}L, \tag{12}$$

where $\hat{X}$ denotes the estimation of the unknown corrections. The estimation of the model parameters can then be obtained by combining the approximations and the corrections. However, during the computation, we find that matrix $A$ is seriously ill-conditioned. It is well known that an ill-posed problem is a great hindrance to obtaining accurate parameter estimation from observation equations. For an ill-posed equation, a small amount of noise in the observations can often bring large uncertainties to the estimation. The detrimental effect of the ill-posed problem is reflected in the large variance of the least-squares estimations. The variance-covariance matrix of the least-squares estimations can be given by [23]:

$$\text{Cov}_{\hat{X}} = \sigma_0^2 \left(A^{\text{T}}A\right)^{-1}, \tag{13}$$

where $\text{Cov}_{\hat{X}}$ represents the variance-covariance matrix of the estimations, and the diagonal elements of the matrix are the variances of the estimations, $\sigma_0^2$ denotes the unit weight variance. We perform singular value decomposition (SVD) on matrix $A$ [24]:

$$A = USG^{\text{T}}, \tag{14}$$

$$S^{\text{T}} = \begin{bmatrix} \lambda_1 & 0 & 0 & 0 & \cdots & 0 \\ 0 & \ddots & 0 & 0 & \cdots & 0 \\ 0 & 0 & \lambda_n & 0 & \cdots & 0 \end{bmatrix}, \tag{15}$$

where $U$ is a $2m \times 2m$ orthogonal matrix of the left singular vectors, $G$ is an $n \times n$ orthogonal matrix of the right singular vectors, and $n$ denotes the number of unknowns, which is equivalent to $m + 3$, $S$ is the matrix of singular values, and $\lambda$ are non-negative real numbers that are conventionally listed in decreasing order, i.e., $\lambda_1 > \lambda_2 > \cdots > \lambda_n$. Using the trace operation, the sum of the variances of the estimations can be given by:

$$\text{Vars}_{\hat{X}} = \text{Trace}(\text{Cov}_{\hat{X}}) = \sigma_0^2 \left(\sum_{i=1}^{n} \frac{1}{\lambda_i^2}\right), \tag{16}$$

where $\text{Vars}_{\hat{X}}$ represents the sum of the variances of the estimations, and Trace denotes the trace operation. If the equation is ill-posed, the singular values gradually decrease to zero, and $\lambda_1$ is much larger than $\lambda_n$ ($\lambda_n$ is close to zero). Equation (16) shows that small singular values greatly magnify the estimation variances. Thus, the least-squares estimation becomes highly unreliable and is unable to obtain accurate estimations of the parameters.

In order to overcome the ill-posedness of the problem, truncated singular value decomposition (TSVD) [26–28] is adopted. TSVD truncates the small singular values that greatly enlarge the variances to improve the least-squares estimation. Using the TSVD-based method, the estimation of the unknown parameters is given by:

$$\hat{X}_t = GS_t^T U^T L, \tag{17}$$

$$S_t^T = \begin{bmatrix} S_P^{-1} & 0 & \cdots & 0 \\ 0 & S_T = \mathbf{0} & \cdots & 0 \end{bmatrix}, \tag{18}$$

where $\hat{X}_t$ denotes the improved estimation by TSVD, $S_t$ denotes the inverse singular value matrix which truncates the small singular values, $S_P$ is the singular value matrix which represents the preserved large singular values, and $S_T$ is a zero matrix that represents the truncated small singular values. We can then obtain the variance of the TSVD estimations as follows:

$$\text{Vars}_{\hat{X}_t} = \text{Trace}\left( GS_t^T U^T \sigma_0^2 US_t G^T \right) = \sigma_0^2 \left( \sum_{i=1}^{k} \frac{1}{\lambda_i^2} \right), \tag{19}$$

where $\text{Vars}_{\hat{X}_t}$ represents the variance of the TSVD estimations. $k$ denotes the number of preserved large singular values. Equation (19) shows that TSVD greatly reduces the variance of the least-squares estimations through truncating the small singular values. However, it is well known that TSVD results in a biased estimation. Truncating the small singular values not only reduces the variance but also introduces bias into the estimation. TSVD improves the estimation by reducing the mean-square error (MSE) of the least-squares estimation. The MSE is expressed as [26]:

$$\text{Mse}_{\hat{X}_t} = \text{Vars}_{\hat{X}_t} + Bias_{\hat{X}_t}^T Bias_{\hat{X}_t}, \tag{20}$$

where $\text{Mse}_{\hat{X}_t}$ represents the MSE of the TSVD estimation, and $Bias_{\hat{X}_t}$ represents the bias introduced by TSVD. The sum of squares of $Bias_{\hat{X}_t}$ is given by:

$$Bias_{\hat{X}_t}^T Bias_{\hat{X}_t} = \sum_{i=k+1}^{n} X^T G_i G_i^T X, \tag{21}$$

where $X$ represents the true values of the unknown parameters, and $G_i$ denotes the $i$-th right singular vector that corresponds to the $i$-th singular value. Equation (21) shows that the bias is introduced by truncating the small singular values. The more singular values that are truncated, the more bias is introduced. Considering Equation (20), it is clear that TSVD reduces the MSE by truncating the small singular values. However, the reduction of the MSE relies on the reduced variance being more than the introduced bias. Therefore, the truncation parameter k which determines the preserved large singular values and truncated small singular values is a key factor for TSVD to reduce the MSE.

The method that is commonly used for the determination of the truncation parameter is related to the condition number, i.e., $\lambda_1 / \lambda_i$, as defined in Equation (15). If the condition number is larger than the given upper limit, the singular value $\lambda_i$ should be truncated [28–30]. However, large-scale PolInSAR data usually consist of millions of pixels, and the singular values in ill-conditioned matrices of pixels are different from each other. Therefore, it is difficult to determine a reasonable upper limit for the condition number. In order to determine a reasonable truncation parameter for this issue, we need to develop a new approach. From Equations (19) and (21), it can be seen that if singular value $\lambda_i$ is truncated, the reduced variances are $\sigma_0^2 / \lambda_i^2$ and the introduced bias is $X^T G_i G_i^T X$. Therefore, if $\sigma_0^2 / \lambda_i^2 > X^T G_i G_i^T X$, the singular value should be truncated, and if $\sigma_0^2 / \lambda_i^2 < X^T G_i G_i^T X$, the singular value should be preserved. In order to determine the small singular values, the values of

$\sigma_0^2/\lambda_i^2$ and $X^T G_i G_i^T X$ need to be calculated accurately. From the least-squares estimation, the estimation of is given by [23]:

$$\hat{\sigma}_0^2 = \frac{V^T V}{2m - n} = \frac{(UU^T L - L)^T (UU^T L - L)}{2m - n},$$ (22)

where $U$ denotes the $2m \times n$ dimensional left singular vectors, $\hat{\sigma}_0^2$ is the estimation of $\sigma_0^2$ [23,26], and $n$ is the number of unknowns. It is clear that the small singular values have no adverse effect on the estimation of $\sigma_0^2$. Therefore, $\sigma_0^2/\lambda_i^2$ can be calculated by the least-squares solution if $2m > n$ [23,26]. In this paper, 10 polarizations are selected for the estimation of the pure volume coherence. Therefore, $\sigma_0^2$ can be estimated.

Using the SVD matrices of $A$, the estimation of $G_i^T X$ by least squares can be expressed as:

$$G_i^T \hat{X} = \lambda_i^{-1} U_i^T L.$$ (23)

It can be seen from Equation (23) that the singular values greatly affect the estimation of $G_i^T X$. Small singular values seriously enlarge the value of $G_i^T \hat{X}$, whereas large singular values do not. The variance of the estimation of $G_i^T X$ can be given by:

$$\mathrm{Var_g} = \lambda_i^{-1} U_i^T \sigma_0^2 U_i \lambda_i^{-1} = \sigma_0^2 \lambda_i^{-2},$$ (24)

where $\mathrm{Var_g}$ represents the variance of estimation $G_i^T \hat{X}$. The equation shows that the variance of $G_i^T \hat{X}$ is negatively correlated with the singular value. Therefore, the variance of $G_i^T \hat{X}$ which corresponds to a large singular value is small. Furthermore, the estimation of $G_i^T X$ is reliable due to the small variance. However, the estimation of $G_i^T X$, which corresponds to a small singular value, is unreliable due to the large variance.

Generally, if the standard deviation of $G_i^T \hat{X}$ is less than $3\hat{\sigma}_0$, the estimation is considered to be reliable. Therefore, the reliable estimations with a standard deviation of less than $3\hat{\sigma}_0$ can be given by $J = \left[ \hat{X}^T G_1 G_1^T \hat{X}, \ \hat{X}^T G_2 G_2^T \hat{X}, \cdots, \hat{X}^T G_j G_j^T \hat{X} \right]$. Since the values of the reduced variances $\hat{\sigma}_0^2/\lambda_i^2$ change as $\hat{\sigma}_0^2/\lambda_1^2 < \hat{\sigma}_0^2/\lambda_2^2 < \cdots < \hat{\sigma}_0^2/\lambda_n^2$, if $\hat{\sigma}_0^2/\lambda_i^2$ is bigger than 90% of the values of $J$, it can be considered that $\sigma_0^2/\lambda_i^2 > X^T G_i G_i^T X$, and singular value $\lambda_i$ needs to be truncated. The other small singular values $\lambda_r$ that are smaller than $\lambda_i$ can also be denoted as $\sigma_0^2/\lambda_r^2 > X^T G_r G_r^T X$ and need to be truncated. The small singular values which should be truncated in the TSVD-based method are thus determined. The ill-posed problem can be well solved by TSVD with the proposed truncation method. Finally, the pure volume coherence can be estimated by the proposed TSVD-based method.

### 3.2. Extraction of Vegetation Height from the Pure Volume Coherence

Since the pure volume coherence is parameterized by $\gamma_v = a + bi$, the pure volume coherence is estimated as $\hat{\gamma}_v = \hat{a} + \hat{b}i$ by the proposed TSVD-based method. Using the estimated parameters $\hat{a}$ and $\hat{b}$, the pure volume coherence which links to the vegetation height can be expressed as:

$$\hat{a} + \hat{b}i = \frac{2\sigma \left( e^{(2\sigma h_v / \cos\theta + ik_z h_v)} - 1 \right)}{(2\sigma + ik_z \cos\theta) \left( e^{(2\sigma h_v / \cos\theta)} - 1 \right)}.$$ (25)

The equation can then be separated into a real part and an imaginary part:

$$\begin{cases} \hat{a} = \mathrm{Re}(\gamma_v), \\ \hat{b} = \mathrm{Im}(\gamma_v). \end{cases}$$ (26)

From Equation (25), it can be seen that $\theta$ and $k_z$ are the known parameters, and the unknown parameters are $\sigma$ and $h_v$. Therefore, $\sigma$ and $h_v$ can be estimated by the least-squares estimation from

Equation (26). The Taylor series is used to convert the nonlinear function into a linear function. The residual functions for the least-squares criterion can then be expressed as [23]:

$$\begin{cases} V_{\text{Re}}^{\text{P}} = \frac{\partial \text{Re}(\gamma_v)}{\partial \sigma} d\sigma + \frac{\partial \text{Re}(\gamma_v)}{\partial h_v} dh_v - l_{\text{Re}}^{\text{P}}, \\ V_{\text{Im}}^{\text{P}} = \frac{\partial \text{Im}(\gamma_v)}{\partial \sigma} d\sigma + \frac{\partial \text{Im}(\gamma_v)}{\partial h_v} dh_v - l_{\text{Im}}^{\text{P}}, \end{cases} \tag{27}$$

where $V_{\text{Re}}^{\text{P}}$ denotes the residual of the real part, and $V_{\text{Im}}^{\text{P}}$ denotes the residual of the imaginary part, $d\sigma$ and $dh_v$ denote the corrections of the approximations of the unknown parameters, and $l_{\text{Re}}^{\text{P}}$ and $l_{\text{Im}}^{\text{P}}$ are the real and imaginary differences between the estimated pure volume coherence and the predicted initial value. Using the vector $V_{\text{p}}$ to denote the residuals, $X_{\text{p}}$ denotes the unknown corrections, $A_{\text{p}}$ denotes the coefficient matrix, and $L_{\text{p}}$ denotes the real and imaginary differences. The residual functions for the least-squares criterion can then be expressed as [23,24,29]:

$$V_{\text{p}} = A_{\text{p}} X_{\text{p}} - L_{\text{p}}. \tag{28}$$

Since the number of residual equations is the same as that of the unknowns, the least-squares criterion can be used to estimate the unknowns. The estimation is given by:

$$\hat{X}_{\text{p}} = \left( A_{\text{p}}^{\text{T}} A_{\text{p}} \right)^{-1} A_{\text{p}}^{\text{T}} L_{\text{p}}, \tag{29}$$

where $\hat{X}_{\text{p}}$ denotes the estimation of the unknowns. The vegetation height can then be obtained from the estimation of $dh_v$ and the initial value of $h_v$ by:

$$\hat{h}_v = h_{v0} + dh_v, \tag{30}$$

where $\hat{h}_v$ denotes the estimated vegetation height, and $h_{v0}$ denotes the initial value of $h_v$.

*3.3. The Determination of Initial Values of Model Parameters*

From Equations (10) and (27), it can be seen that the proposed method is plagued by the initial value. If the initial value cannot be well determined, it is difficult to get reliable estimation. The classical six-dimensional nonlinear optimization method [6] is also confronted by this problem since it is difficult to obtain the priori information of the forest parameters (forest height and extinction). However, it is easy to obtain reliable initial values of the ground phase, the pure volume coherence and the ground-to-volume ratio according to the RVoG assumption and its linear geometrical expression in the complex plane [10]. Based on this, Equation (10) is used to estimate the pure volume coherence, which is important to invert the forest parameters. Compared to the six-dimensional nonlinear optimization method, this method can avoid significant biases caused by the unreliable initial values of forest parameters. Moreover, in comparison to the three-stage method, the proposed method can provide more accurate pure volume coherence because it is free from the assumption that there is one polarization whose ground-to-volume power ratio should be less than −10 dB [10], which cannot be fulfilled for the low-frequency PolInSAR data or for the sparse forest [31]. In addition, only one polarimetric observation is used to calculate the pure volume coherence in the three-stage method. However, multi-polarization observations are used to estimate the pure volume coherence under the TSVD based least-squares estimation framework, which enhances the method's ability to alleviate the effect of observational errors in the estimations. As a result, the obtained pure volume coherence can support more accurate forest parameter estimation.

Then, with the estimated pure volume coherence, Equation (27) is used to estimate forest height. The initial values for the forest height and the mean extinction coefficient are determined by the three-stage method. Although the three-stage method cannot give satisfactory results, especially for the low-frequency PolInSAR data or for sparse forest, these estimated results can be regarded as the initial values for Equation (27). The least-squares method estimates the corrections of initial values from the reliable pure volume coherence to improve the accuracy of the estimated unknown parameters. The final experiment has shown that with the accurate pure volume coherence and reasonable initial values, the proposed method can attain good forest height results.

## 4. Examples

### 4.1. Simulated Experiments

In order to evaluate the performance of the proposed TSVD-based inversion method for the estimation of vegetation height from single-baseline PolInSAR data, we simulated single-baseline PolInSAR data through the PolSARpro tool released by European Space Agency (ESA), Rome, Italy, using the following forest scenario, as shown in Table 1: simulated broad-leaved forest with an average height of 18 m; and ground phase of 0 degrees, corresponding to a ground elevation of 0 m. The complex interferometric coherence of the linear-basis polarizations (transmitted polarization and received polarization are horizontal polarizations, HH, transmitted polarization and received polarization are vertical polarizations, VV) and the Pauli-basis polarization (transmitted polarization is horizontal polarization and received polarization is vertical polarization, HV) could then be obtained. The simulated forest area and the Pauli RGB composite image are shown in Figure 2.

**Table 1.** Parameters of the simulated scenario.

| Platform Height (m) | Center Frequency (Hz) | Incidence Angle (D) | Vertical Baseline (m) | Horizontal Baseline (m) | Vertical Wave Number | Ground Phase (D) | Forest Height (m) |
|---|---|---|---|---|---|---|---|
| 3000 | 1.3 G | 45 | 1 | 10 | 0.1154 | 0 | 18 |

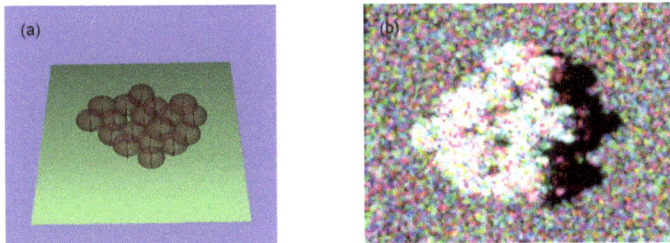

**Figure 2.** (a) simulated forest area; (b) Pauli RGB composite image.

Furthermore, using Pauli-basis polarization, magnitude diversity optimization polarization [6], and phase diversity polarization [25], the complex interferometric coherence observations of HH+VV, HH-VV, opt1, opt2, opt3, phase diversity (PD)-high, and phase diversity (PD)-low could be obtained. Therefore, 20 residual equations could be established from the above 10 complex interferometric coherence observations based on Equation (10) with 13 unknown parameters. The unknown parameters were then estimated by the TSVD-based method. From the estimated parameters and the pure volume coherence model which links to the forest height, the forest height could be extracted by Equation (29). The extraction results are shown in Figure 3b. For comparison, the extraction results of the three-stage method are shown in Figure 3a.

From Figure 3, it can be seen that the TSVD-based method performs much better than the three-stage method in forest height inversion in this test. Clearly, from Figure 3a, the three-stage method fails to invert the forest heights of the rectangular areas of the figure, but the TSVD-based method effectively improves the inversion of the rectangular areas, as can be seen in Figure 3b. The mean values of the extracted forest heights by the three-stage method and the TSVD-based method are 13.383 m and 15.4440 m. This demonstrates that the forest height estimated by TSVD is more accurate than that estimated by the three-stage method. For a further comparison, 16 forest stands were selected from the simulated forest area, and the root-mean-square error (RMSE) of each stand was adopted to compare the performance of the two methods. The RMSEs of each stand are shown in Figure 4.

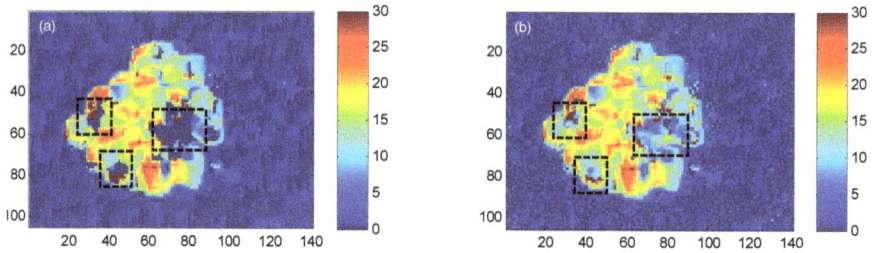

**Figure 3.** (**a**) forest height inversion result of the three-stage method; and (**b**) forest height inversion result of the truncated singular value decomposition (TSVD)-based inversion method.

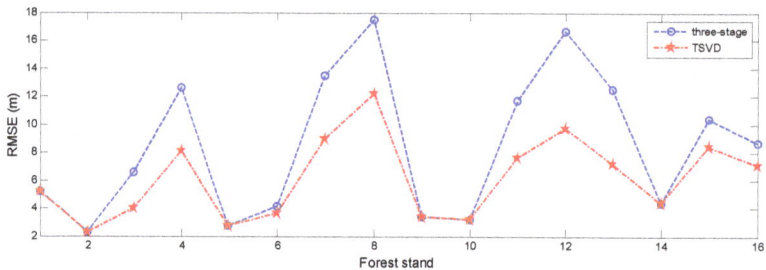

**Figure 4.** Root-mean-square errors (RMSEs) of each forest stand.

For the 16 forest stands, the RMSE of the TSVD-based method is consistently less than that of the three-stage method. This indicates that the inverted forest height obtained by TSVD is closer to the true height than the height obtained by the three-stage method. This further validates the performance of the TSVD-based method.

### 4.2. Validation with E-SAR P-Band Data

#### 4.2.1. Study Area and Data Sets

The proposed TSVD-based method was also applied to E-SAR P-band PolInSAR data, which were collected under the framework of the BioSAR 2008 campaign by the German Aerospace Center, Munich, Germany. The test site is a forest area within the Krycklan River catchment in Northern Sweden, and is mainly covered by mixed boreal forest with heights ranging from 0 to 35 m. The topography elevation is between 150 to 380 m above mean sea level (AMSL). The baseline PolInSAR data were acquired in the repeat-pass configuration. Moreover, as part of the BioSAR2008 campaign, a light detection and ranging (LiDAR) measurement was also obtained by the Swedish Defense Research Agency (FOI). The derived forest height is regarded as the reference in the following analysis.

In this experiment, the forest heights were extracted from single-baseline data. The temporal and spatial baselines were 70 min and 32 m, respectively. The vertical wavenumber ranged from 0.051 to 0.181.The Pauli-basis RGB composite intensity image for the test site is shown in Figure 5.

**Figure 5.** Test site Pauli-basis RGB composite intensity image.

4.2.2. Forest Height Inversion

Following the steps of the TSVD-based inversion method, the complex interferometric coherence observations of HH, VV, HV, HH+VV, HH-VV, opt1, opt2, opt3, PDhigh, and PDlow were used to establish the residual equations. The pure volume coherence was then estimated by the TSVD-based method from the residual equations and, finally, the forest heights were extracted from the pure volume coherence model using the estimated model parameters. The extracted forest heights are shown in Figure 6b. As in the simulated experiments, the three-stage method was also used to extract the forest heights, and the inversion results are shown in Figure 6a. Figure 6c shows the LiDAR forest heights used as a reference.

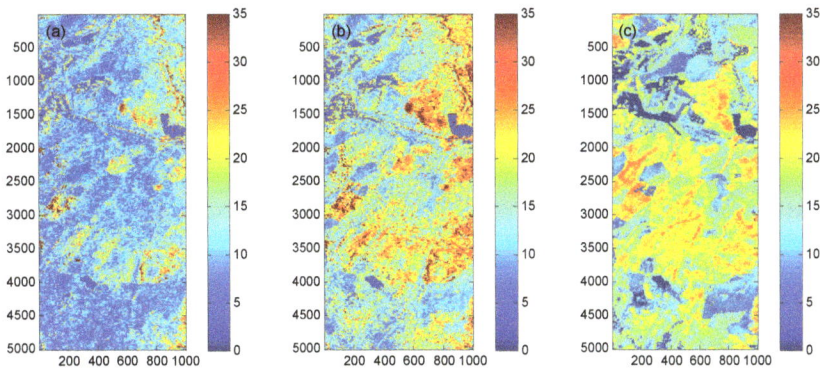

**Figure 6.** (**a**) inversion results of the three-stage method; (**b**) inversion results of the TSVD-based method; (**c**) forest heights derived by light detection and ranging (LiDAR).

It can be seen from Figure 6 that the inversion results of the three-stage method and the TSVD-based method follow a similar spatial trend, but significant differences are also apparent. Compared to the LiDAR results, the inverted forest heights obtained by TSVD are clearly more accurate than those of the three-stage method. In order to analyze the differences, 272 forest stands characterized by nearly uniform tree heights were selected from the LiDAR results. We took the estimated forest height average for every stand and computed the difference between that and the LiDAR forest height. The RMSE and correlation coefficient ($R^2$) were calculated to validate the performance. The validated stand-level plots are displayed in Figure 7.

The validated plots of the three-stage method and the TSVD-based method are characterized by $R^2$ values of 0.2166 and 0.5824, respectively. This indicates that the forest heights inverted by TSVD are closer to the LiDAR forest heights. The RMSEs of the three-stage method and TSVD are 6.6351 and 3.4096, respectively. Clearly, the inversion accuracy of TSVD is higher than that of the three-stage method, showing an improvement of 48.6%. Therefore, it is possible to state that the TSVD-based method can improve the inversion of the forest height in this test site.

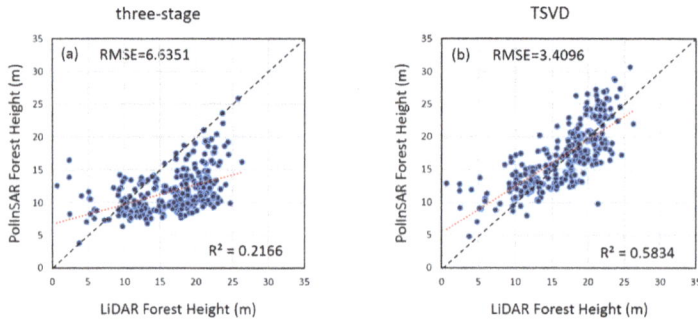

**Figure 7.** (**a**) forest heights estimated by the three-stage method; (**b**) forest heights estimated by the TSVD-based method.

## 5. Discussion

### 5.1. The Extracted Ground Surface Phases by the Three-Stage Method and TSVD

From Equation (10), it can be seen that the ground surface phase can also be estimated by the TSVD-based method. Since the ground surface phase plays an important part in the estimation of the underlying topography [30,32,33], the extracted ground surface phases obtained by the three-stage method and TSVD are shown in Figure 8a,b, respectively.

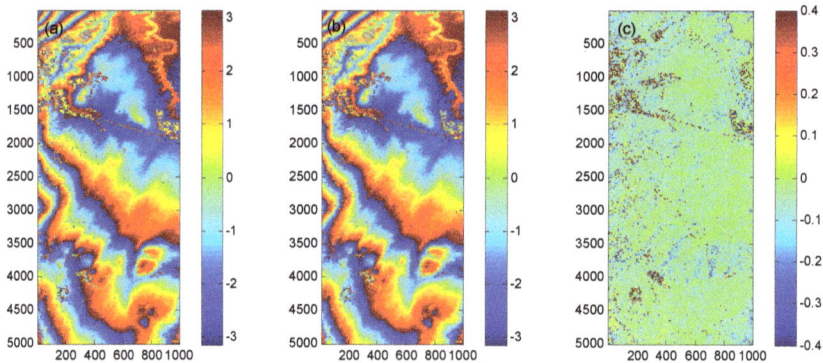

**Figure 8.** (**a**) ground surface phase estimated by the three-stage method; (**b**) ground surface phase estimated by TSVD; (**c**) the difference between the ground surface phases obtained by the three-stage method and TSVD.

It is difficult to see any difference between Figure 8a,b. This indicates that the three-stage method and TSVD perform similarly in extracting the ground surface phase. For a more in-depth analysis, the ground surface phase obtained by TSVD subtracted from that obtained by the three-stage method is shown in Figure 8c. It can be seen that most of the values in Figure 8c are close to zero. This further validates that TSVD is unable to improve the accuracy of the estimation of the ground surface phase and the underlying digital elevation model (DEM) [30].

The line-fit approach is used in the three-stage method to determine the ground surface phase. In order to reconstruct the straight line accurately from noisy coherence sets, 10 polarizations are used by least-squares based line-fit approach. Similarly, the same 10 polarizations are used in TSVD for the estimation of model parameters. Therefore, the observations used by line-fit approach and TSVD have the same noise and information, and the basic criterion in the line-fit approach and TSVD both are least-squares. This is a possible reason for this result.

## 5.2. Effects on Estimation of Phase Height

As mentioned in Section 3, the TSVD-based method first estimates the pure volume coherence from the complex interferometric coherence and then extracts the forest height. The phase height, which is very important for the inversion of forest height [34,35], was also computed from the estimated pure volume coherence and compared with the result of the three-stage method. The phase heights calculated from the estimated pure volume coherence obtained by the three-stage method and TSVD are shown in Figure 9a,b.

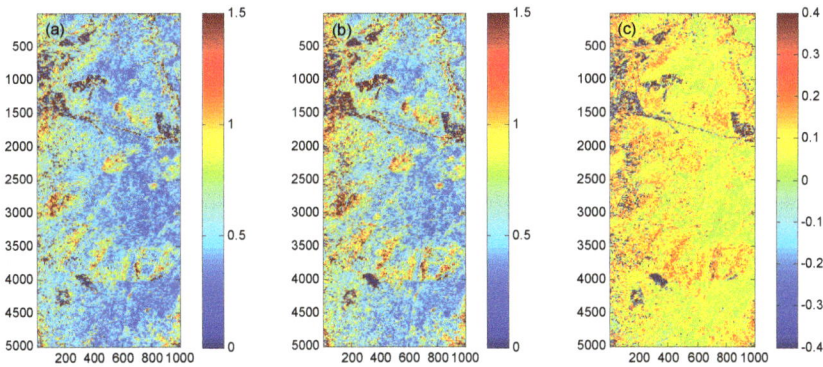

**Figure 9.** (**a**) the phase heights extracted from the pure volume coherence obtained by the three-stage method; (**b**) the phase heights extracted from the pure volume coherence obtained by TSVD; (**c**) the differences between the phase heights obtained by the three-stage method and TSVD.

From Figure 9a,b, it can be seen that the estimated phase heights in Figure 9a are higher than those in Figure 9b. We subtracted the phase heights in Figure 9b from those in Figure 9a and display the results in Figure 9c. Clearly, most of the values in Figure 9c are positive numbers. This indicates that the phase height extracted by TSVD is a better fit for the theoretical phase height of the pure volume coherence than the phase height extracted by the three-stage method. For a further comparison, we computed the average phase height of each forest stand and show the results in Figure 10.

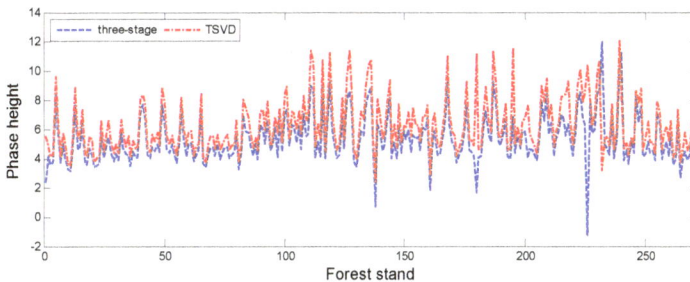

**Figure 10.** The estimated phase heights for each forest stand.

Figure 10 clearly indicates that the phase heights estimated by the TSVD-based method are consistently higher than those estimated by the three-stage method in each forest stand. Therefore, the forest heights inverted by TSVD are more accurate than those inverted by the three-stage method in the forest stands. To confirm this conclusion, we also used the least-squares method to extract the forest heights from the pure volume coherence estimated by the three-stage method, as mentioned in Section 3.2. The results are shown in Figure 11.

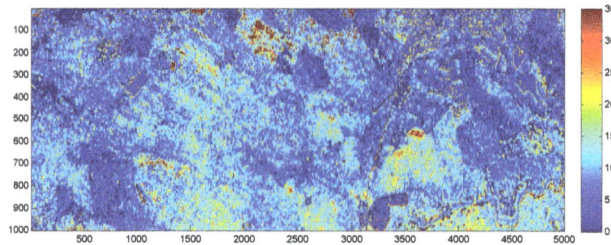

**Figure 11.** Forest heights extracted by the least-squares method from the pure volume coherence estimated by the three-stage method.

Comparing Figures 11 and 5b, which displays the inversion results of TSVD, clearly, the forest heights in Figure 5b are closer to the LiDAR forest heights than those in Figure 11. Therefore, we can conclude that the TSVD-based method has the capacity to improve the estimation of the pure volume coherence. Based on the improved pure volume coherence, the forest height can be extracted more accurately.

*5.3. Limitations of the TSVD-Based Method*

Two problems are worth discussing. Firstly, TSVD plays an important role in the proposed solution. Since the ordinary truncation method is not suitable for this issue, a more adaptive truncation method is proposed in this paper. From Equation (23), it can be seen that sufficient polarizations are needed to estimate $\sigma_0^2$, i.e., $2m > n$ [23,26]. In this paper, 10 polarizations are selected for the forest height inversion. Therefore, $2m - n = 7$ confirms the accuracy of the estimation of $\sigma_0^2$ [23]. Secondly, due to the scattering mechanism, the inverted forest heights always follow a spatial trend, which can be seen in the results of both the three-stage method and TSVD. As a consequence, it is apparent that the far-range areas result in the overestimation of the forest height in Figure 6a,b, and especially in Figure 6b.

# 6. Conclusions

A TSVD-based method has been proposed in this paper for forest height inversion from single-baseline PolInSAR data. Differing from the traditional three-stage method, the new method estimates the pure volume coherence intuitively from the complex interferometric coherence, and has the capacity to adjust the contributions of the polarizations in the estimation of the model parameters. The TSVD-based method was first applied in forest height inversion from simulated PolInSAR data generated in PolSARpro. The results demonstrated that the TSVD-based method significantly improves the inversion results when compared to the three-stage method. This was also confirmed with airborne E-SAR P-band data obtained over a mixed boreal forest. The inverted forest heights obtained by TSVD showed an improvement in RMSE of 48.6% when compared to the results of the three-stage method. The phase heights of the estimated pure volume coherence were also well improved when compared to the results of the three-stage method.

**Acknowledgments:** The work presented in the paper was supported by the Nature Science Foundation of China (Nos. 41531068, 41474008, 41574006, 41674012). The work also was supported by Hunan Provincial Department of Education Science Research Key Project (No. 15A074).

**Author Contributions:** Dongfang Lin conceived the idea, designed and performed the experiments, and wrote and revised the paper; Jianjun Zhu analyzed the PolInSAR experiments and revised the paper; Haiqiang Fu analyzed the experimental results and revised the paper; Qinghua Xie analyzed the experimental results; and Bing Zhang performed the simulated experiments.

**Conflicts of Interest:** The authors declare no conflict of interest. The founding sponsors had no role in the design of the study; in the collection, analyses, or interpretation of data; in the writing of the manuscript, and in the decision to publish the results.

## References

1.  Balzter, H.; Rowland, C.S.; Saich, P. Forest canopy height and carbon estimation at Monks Wood National Nature Reserve, UK, using dual-wavelength SAR interferometry. *Remote Sens. Environ.* **2007**, *108*, 224–239. [CrossRef]
2.  Gama, F.F.; dos Santos, J.R.; Mura, J.C. Eucalyptus biomass and volume estimation using interferometric and polarimetric SAR data. *Remote Sens.* **2010**, *2*, 939–956. [CrossRef]
3.  Luo, H.M.; Li, X.W.; Chen, E.; Cheng, J.; Cao, C. Analysis of forest backscattering characteristics based on polarization coherence tomography. *Sci. China Technol. Sci.* **2011**, *53*, 166–175. [CrossRef]
4.  Neumann, M.; Ferro-Famil, L.; Reigber, A. Estimation of forest structure, ground, and canopy layer characteristics from multibaseline polarimetric interferometric SAR data. *IEEE Trans. Geosci. Remote Sens.* **2010**, *48*, 1086–1104. [CrossRef]
5.  Cloude, S.R.; Papathanassiou, K.P. Polarimetric SAR interferometry. *IEEE Trans. Geosci. Remote Sens.* **1998**, *36*, 1551–1565. [CrossRef]
6.  Papathanassiou, K.P.; Cloude, S.R. Single-baseline polarimetric SAR interferometry. *IEEE Trans. Geosci. Remote Sens.* **2001**, *39*, 2352–2363. [CrossRef]
7.  Kugler, F.; Schulze, D.; Hajnsek, I.; Pretzsch, H.; Papathanassiou, K.P. Tan DEM-X Pol-InSAR performance for forest height estimation. *IEEE Trans. Geosci. Remote Sens.* **2014**, *52*, 6404–6422. [CrossRef]
8.  Garestier, F.; Dubois-Fernandez, P.C.; Papathanassiou, K.P. Pine forest height inversion using single-pass X-band PolInSAR data. *IEEE Trans. Geosci. Remote Sens.* **2008**, *46*, 56–68. [CrossRef]
9.  Garestier, F.; Toan, T.L. Forest modeling for height inversion using single baseline InSAR/PolIn SAR data. *IEEE Trans. Geosci. Remote Sens.* **2010**, *48*, 1528–1539. [CrossRef]
10. Cloude, S.R. *Polarisation: Applications in Remote Sensing*; Oxford University Press: New York, NY, USA, 2009.
11. Treuhaft, R.N.; Madsen, S.N.; Moghaddam, M.; Zyl, J.J.V. Vegetation characteristics and underlying topography from interferometric data. *Radio Sci.* **1996**, *31*, 1449–1495. [CrossRef]
12. Treuhaft, R.N.; Siqueira, P.R. Vertical structure of vegetated land surfaces from interferometric and polarimetric data. *Radio Sci.* **2000**, *35*, 141–177. [CrossRef]
13. Garestier, F.; Toan, T.L. Estimation of the backscatter vertical profile of a pine forest using single baseline P-band (Pol-) InSAR data. *IEEE Trans. Geosci. Remote Sens.* **2010**, *48*, 3340–3348. [CrossRef]
14. Garestier, F.; Dubois-Fernandez, P.C.; Champion, I. Forest height inversion using high resolution P-band Pol-InSAR data. *IEEE Trans. Geosci. Remote Sens.* **2008**, *46*, 3544–3559. [CrossRef]
15. Parks, J.; Kugler, F.; Papathanassiou, K.P.; Hajnsek, I.; Hallikainen, M. Height estimation of boreal forest: Interferometric model-based inversion at L- and X-band versus HUTSCAT profiling scatterometer. *IEEE Geosci. Remote Sens. Lett.* **2007**, *4*, 466–470. [CrossRef]
16. Li, X.W.; Guo, H.D.; Liao, J.J.; Wang, C.L.; Yan, F.L. Inversion of vegetation parameters using spaceborne polarimetric SAR interferometry. *J. Remote Sens.* **2002**, *6*, 424–429.
17. Ballester-Berman, J.D.; Vicente-Guijalba, F.; Lopez-Sanchez, J.M. A simple RVoG test for PolInSAR data. *IEEE J. Sel. Top. Appl. Earth Obs. Remote Sens.* **2015**, *8*, 1028–1040. [CrossRef]
18. Cloude, S.R.; Papathanassiou, K.P. Three-stage inversion process for polarimetric SAR interferometry. *IEEE Proc. Radar Sonar Navig.* **2003**, *150*, 125–134. [CrossRef]
19. Chen, E.X.; Li, Z.Y.; Pang, Y.; Tian, X. Polarimetric synthetic aperture radar interferometry based mean tree height extraction technique. *Sci. Silvae Sin.* **2007**, *43*, 66–70.
20. Fu, H.Q.; Wang, C.C.; Zhu, J.J.; Xie, Q.H.; Zhao, R. Inversion of forest height from PolInSAR using complex least squares adjustment method. *Sci. China Earth Sci.* **2015**, *58*, 1018–1031. [CrossRef]
21. Small, D. Generation of Digital Elevation Models through Spaceborne SAR Interferometry. Ph.D. Thesis, University of Zurich, Zurich, Switzerland, 1998.

22. Lavalle, M.; Khun, K. Three-baseline InSAR estimation of forest height. *IEEE Geosci. Remote Sens. Lett.* **2014**, *11*, 1737–1741. [CrossRef]
23. Cui, X.; Yu, Z.; Tao, B.; Liu, D.; Yu, Z.; Sun, H.; Wang, X. *Generalized Surveying Adjustment*, 2nd ed.; Wuhan University Press: Wuhan, China, 2009.
24. Lin, D.; Zhu, J.; Song, Y.; He, Y. Construction method of regularization by singular value decomposition of design matrix. *Acta Geod. Cartogr. Sin.* **2016**, *45*, 883–889.
25. Tabb, M.; Orrey, J.; Flynn, T. Phase Diversity: An optimal decomposition for vegetation parameter estimation using polarimetric SAR interferometry. In Proceeding of the 4th European Conference on Synthetic Aperture Radar, Köln, Germany, 2–4 June 2002; pp. 721–724.
26. Xu, P.L. Truncated SVD methods for discrete linear ill-posed problems. *Geophys. J. Int.* **1998**, *135*, 505–514. [CrossRef]
27. Hansen, P.C. The truncated SVD as a method for regularization. *BIT Numer. Math.* **1987**, *27*, 534–553. [CrossRef]
28. Reichel, L.; Rodriguez, G. Old and new parameter choice rules for discrete ill-posed problems. *Numer. Algorithms* **2013**, *63*, 65–87. [CrossRef]
29. Zhu, J.J.; Xie, Q.H.; Zuo, T.Y.; Wang, C.C.; Xie, J. Criterion of complex least squares adjustment and its application in tree inversion with PolInSAR data. *Acta Geod. Cartogr. Sin.* **2014**, *43*, 45–51.
30. Lopez-Martinez, C.; Papathanassiou, K.P. Cancellation of scattering mechanisms in PolInSAR: Application to underlying topography estimation. *IEEE Trans. Geosci. Remote Sens.* **2013**, *51*, 953–965. [CrossRef]
31. Wang, C.C.; Wang, L.; Fu, H.Q.; Xie, Q.H.; Zhu, J. The impact of forest density on forest height inversion modeling from Polarimetric InSAR data. *Remote Sens.* **2016**, *8*, 291. [CrossRef]
32. Tebaldini, S. Single and multipolarimetric SAR tomography of forested areas: A parametric approach. *IEEE Trans. Geosci. Remote Sens.* **2010**, *48*, 2375–2387. [CrossRef]
33. Tebaldini, S.; Rocca, F. Multibaseline polarimetric SAR tomography of a boreal forest at P- and L-bands. *IEEE Trans. Geosci. Remote Sens.* **2012**, *50*, 232–246. [CrossRef]
34. Tebaldini, S. Multi-Baseline SAR Imaging: Models and Algorithms. Ph.D. Thesis, Politecnico Di Milano, Milano, Italy, 11 October 2009.
35. Lee, J.; Hoppel, K.; Mango, S.; Miller, A. Intensity and phase statistics of multilook polarimetric and interferometric SAR image. *IEEE Trans. Geosci. Remote Sens.* **1994**, *32*, 1017–1028.

MDPI AG

St. Alban-Anlage 66

4052 Basel, Switzerland

Tel. +41 61 683 77 34

Fax +41 61 302 89 18

http://www.mdpi.com

*Applied Sciences* Editorial Office

E-mail: applsci@mdpi.com

http://www.mdpi.com/journal/applsci

www.ingramcontent.com/pod-product-compliance
Lightning Source LLC
Chambersburg PA
CBHW041215220326
41597CB00033BA/5971